GCSE 9–1 FOUNDATION MATHEMATICS EDEXCEL EXAM PRACTICE

Naomi Norman

Author Naomi Norman
Editorial team Haremi Ltd
Series designers emc design ltd
Typesetting York Publishing Solutions Pvt. Ltd.
Illustrations York Publishing Solutions Pvt. Ltd.
App development Hannah Barnett, Phil Crothers and Haremi Ltd

Designed using Adobe InDesign
Published by Scholastic Education, an imprint of Scholastic Ltd, Book End, Range Road, Witney, Oxfordshire, OX29 0YD
Registered office: Westfield Road, Southam, Warwickshire CV47 0RA
www.scholastic.co.uk

Printed and bound in India by Replika Press Pvt. Ltd.

1 2 3 4 5 6 7 8 9   7 8 9 0 1 2 3 4 5 6

British Library Cataloguing-in-Publication Data
A catalogue record for this book is available from the British Library.
ISBN 978-1407-16899-9

Notes from the publisher

Please use this product in conjunction with the official specification and sample assessment materials. Ask your teacher if you are unsure where to find them.

The marks and star ratings have been suggested by our subject experts, but they are to be used as a guide only.

Answer space has been provided, but you may need to use additional paper for your workings.

Contents

How to use this book

This Exam Practice Book has been produced to help you revise for your 9–1 GCSE in Edexcel Foundation Mathematics. Written by an expert and packed full of exam-style questions for each subtopic, along with full practice papers, it will get you exam ready!

The best way to retain information is to take an active approach to revision. Don't just read the information you need to remember – do something with it! Transforming information from one form into another and applying your knowledge will ensure that it really sinks in. Throughout this book you'll find lots of features that will make your revision practice an active, successful process.

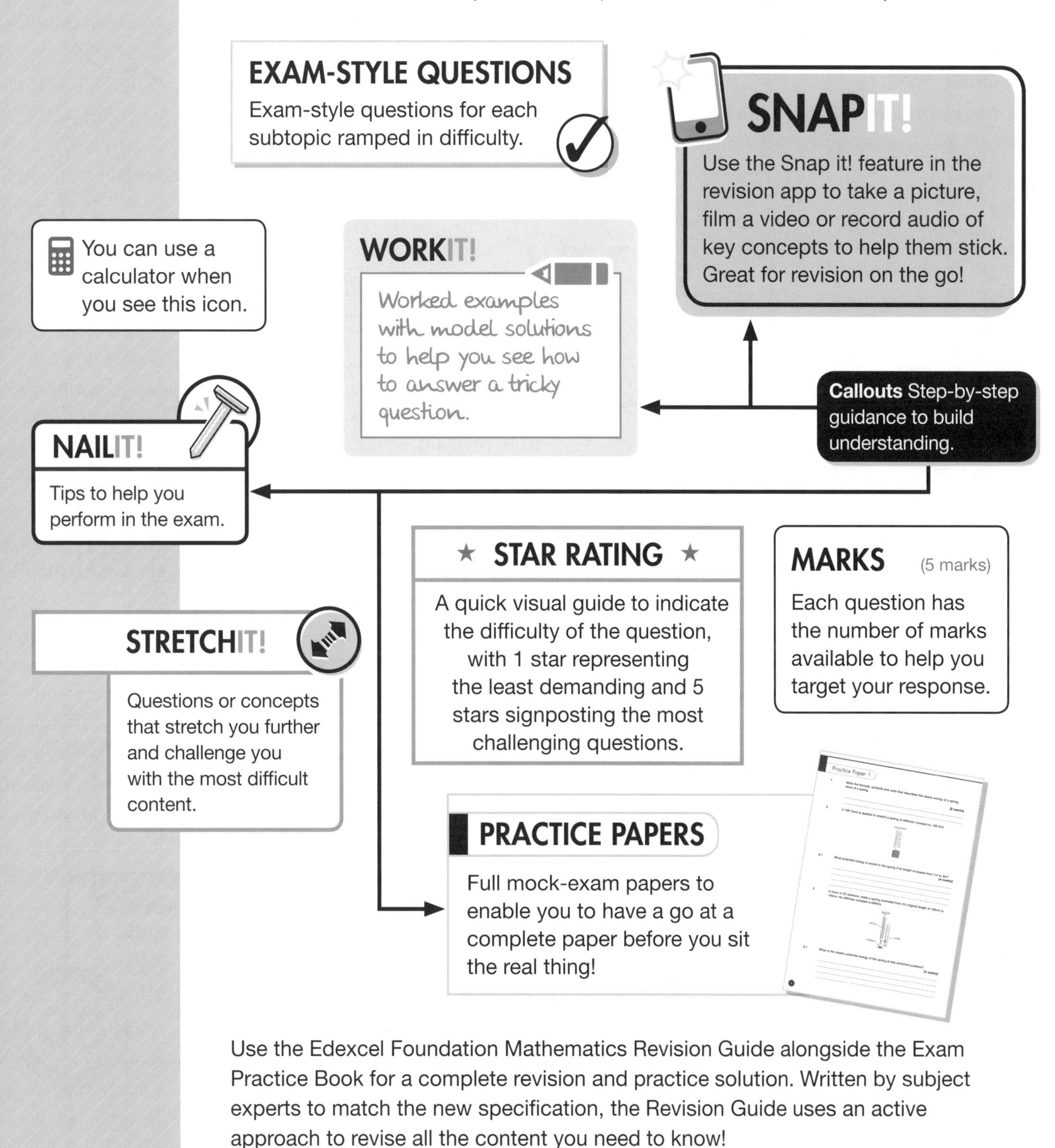

Use the Edexcel Foundation Mathematics Revision Guide alongside the Exam Practice Book for a complete revision and practice solution. Written by subject experts to match the new specification, the Revision Guide uses an active approach to revise all the content you need to know!

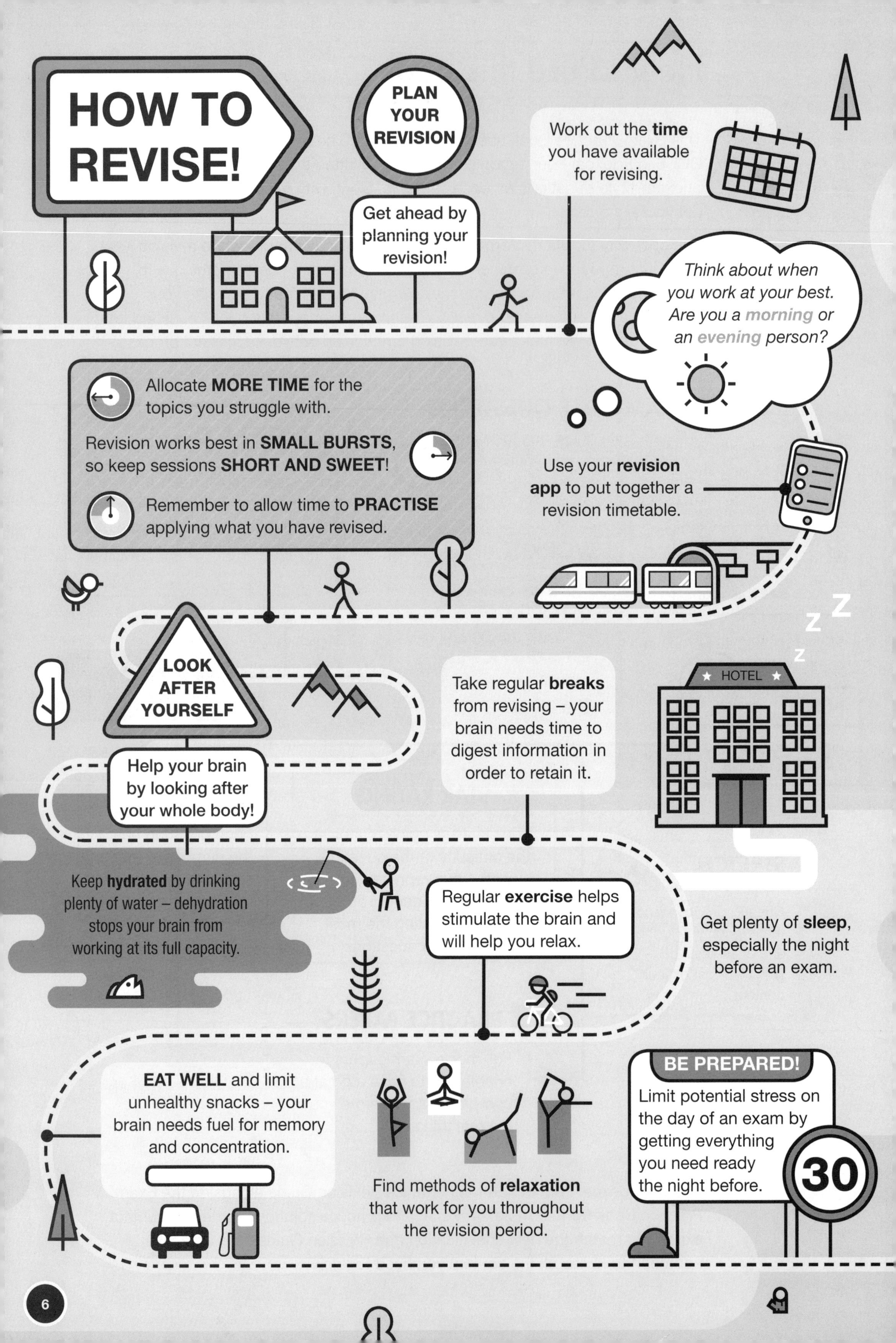
HOW TO REVISE!
PLAN YOUR REVISION
Get ahead by planning your revision!
Work out the **time** you have available for revising.
*Think about when you work at your best. Are you a **morning** or an **evening** person?*
Allocate **MORE TIME** for the topics you struggle with.
Revision works best in **SMALL BURSTS**, so keep sessions **SHORT AND SWEET**!
Remember to allow time to **PRACTISE** applying what you have revised.
Use your **revision app** to put together a revision timetable.
LOOK AFTER YOURSELF
Help your brain by looking after your whole body!
Take regular **breaks** from revising – your brain needs time to digest information in order to retain it.
HOTEL
Keep **hydrated** by drinking plenty of water – dehydration stops your brain from working at its full capacity.
Regular **exercise** helps stimulate the brain and will help you relax.
Get plenty of **sleep**, especially the night before an exam.
EAT WELL and limit unhealthy snacks – your brain needs fuel for memory and concentration.
Find methods of **relaxation** that work for you throughout the revision period.
BE PREPARED!
Limit potential stress on the day of an exam by getting everything you need ready the night before.
30

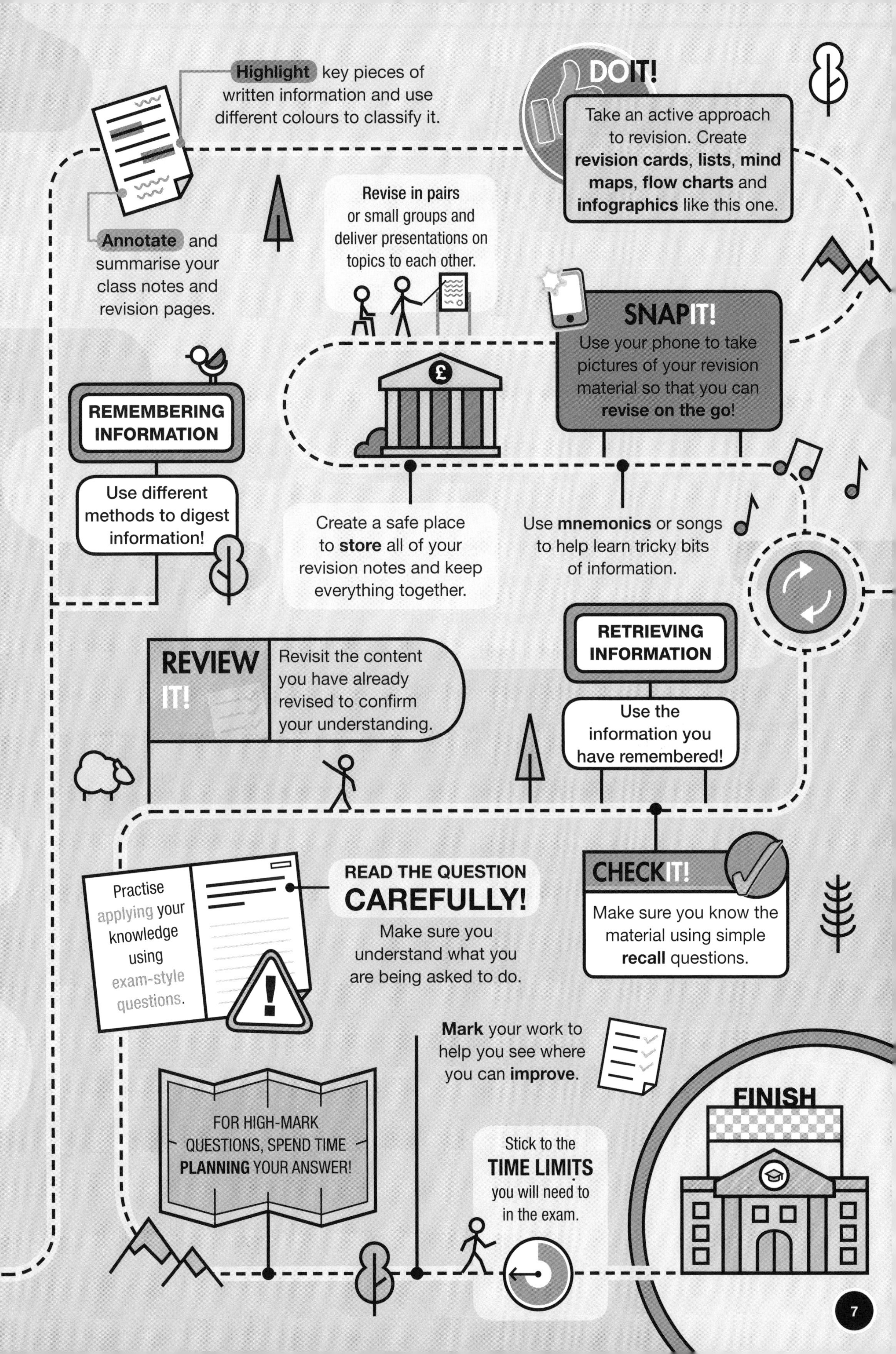
Highlight key pieces of written information and use different colours to classify it.
DO IT!
Take an active approach to revision. Create revision cards, lists, mind maps, flow charts and infographics like this one.
Annotate and summarise your class notes and revision pages.
Revise in pairs or small groups and deliver presentations on topics to each other.
SNAP IT!
Use your phone to take pictures of your revision material so that you can revise on the go!
REMEMBERING INFORMATION
Use different methods to digest information!
Create a safe place to store all of your revision notes and keep everything together.
Use mnemonics or songs to help learn tricky bits of information.
RETRIEVING INFORMATION
Use the information you have remembered!
REVIEW IT!
Revisit the content you have already revised to confirm your understanding.
Practise applying your knowledge using exam-style questions.
READ THE QUESTION CAREFULLY!
Make sure you understand what you are being asked to do.
CHECK IT!
Make sure you know the material using simple recall questions.
Mark your work to help you see where you can improve.
FOR HIGH-MARK QUESTIONS, SPEND TIME PLANNING YOUR ANSWER!
Stick to the TIME LIMITS you will need to in the exam.
FINISH

Number

Factors, multiples and primes

① Find the Highest Common Factor (HCF) of 18 and 24. (2 marks, ★★)

For HCF: List all the factors for each number. Now look for the highest factor in **both** lists.

....

② List all the prime numbers between 15 and 25. (2 marks, ★★)

....

③ Write 60 as a product of its prime factors. (2 marks, ★★★)

Give your answer in index form, like this: 60 = □$^{▪}$ × □ × □

....

④ In a piece of music there are two drummers. .

Drummer 1 hits her drum after 6 seconds.

She then hits her drum every 6 seconds after that.

Drummer 2 hits his drum after 8 seconds.

Drummer 2 hits his drum every 8 seconds after that.

How many times do the drummers hit their drums at the same time in the first minute?

Show working to justify your answer. (3 marks, ★★★★★)

List all the times Drummer 1 will hit her drum in the first minute: 6, 12… seconds. Then list all the times Drummer 2 will hit his drum in the first minute.

....

....

....

....

....

STRETCHIT!

Can you write a factor tree for the number 216?

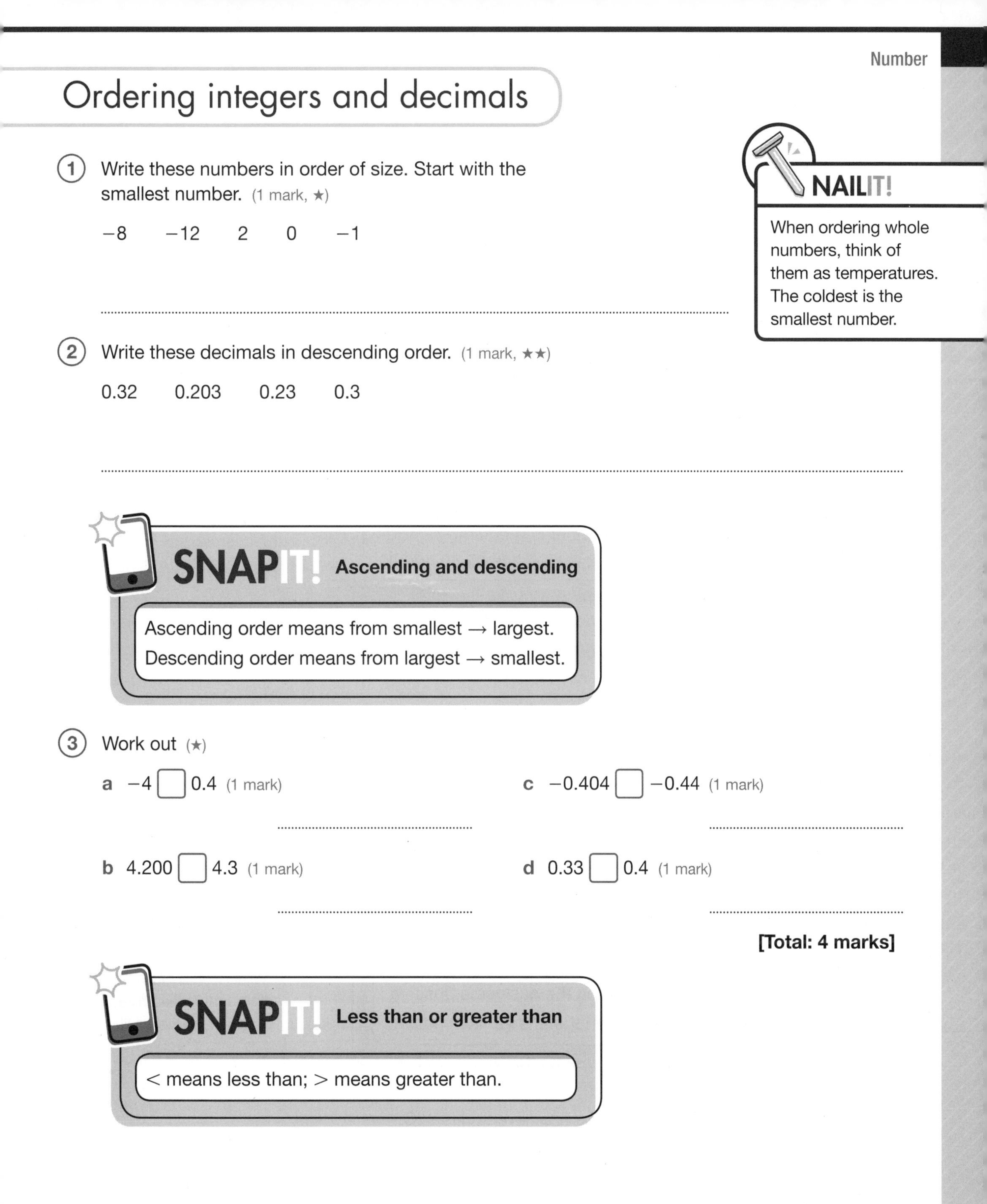

Ordering integers and decimals

NAILIT!

When ordering whole numbers, think of them as temperatures. The coldest is the smallest number.

1. Write these numbers in order of size. Start with the smallest number. (1 mark, ★)

   $-8 \quad -12 \quad 2 \quad 0 \quad -1$

   ....

2. Write these decimals in descending order. (1 mark, ★★)

   0.32 0.203 0.23 0.3

   ....

SNAPIT! **Ascending and descending**

Ascending order means from smallest → largest.
Descending order means from largest → smallest.

3. Work out (★)

   a -4 ☐ 0.4 (1 mark) ....

   b 4.200 ☐ 4.3 (1 mark) ....

   c -0.404 ☐ -0.44 (1 mark) ....

   d 0.33 ☐ 0.4 (1 mark) ....

[Total: 4 marks]

SNAPIT! **Less than or greater than**

$<$ means less than; $>$ means greater than.

Calculating with negative numbers

1. Work out (★)

   a $-7 + -3$ (1 mark) ....

   b $-7 - -3$ (1 mark) ....

   c $8 + -5 - -2$ (1 mark) ....

   d $-4 - -6 + -1$ (1 mark) ....

[Total: 4 marks]

SNAPIT! Adding or subtracting negative numbers

Adding a negative number is the same as subtracting. This means you can rewrite the calculation as a subtraction.

Subtracting a negative number is the same as adding. This means you can rewrite the calculation as an addition.

2. Work out (★)

   a -9×2 (1 mark) ....

   b $-12 \div -3$ (1 mark) ....

   c $-4 \times -2 \times 5$ (1 mark) ....

   d $-24 \div 3 \times 2$ (1 mark) ....

[Total: 4 marks]

SNAPIT! Multiplying or dividing negative numbers

When multiplying and dividing negative numbers

- if the signs are the same, then the answer is positive
- if the signs are different, then the answer is negative.

STRETCHIT!

What would happen if you multiplied three negative numbers like $-2 \times -3 \times -4$?

3. The temperature at 6 pm one December evening is 2 °C.

   By 10 pm it has dropped by 4 °C.

   By 2 am it has dropped by another 5 °C.

   What is the temperature at 2 am? (2 marks, ★★)

   ....

4. There are five questions in a quiz. People score 3 marks for each correct answer and -2 marks for each incorrect answer.

   Sally scores -5. How many correct and incorrect answers did she get? (3 marks, ★★★★)

   ....

Multiplication and division

NAILIT!

Read the question carefully. Decide if you need to multiply or divide.

1 Work out (★)

a 357 × 6 (1 mark)

....

b 261 × 43 (1 mark)

....

c 552 ÷ 6 (1 mark)

....

d 676 ÷ 13 (1 mark)

....

[Total: 4 marks]

2 300 textbooks are packed into boxes. 24 books fit in each box. (★★)

a How many boxes are filled? (2 marks)

....

b How many books are left over? (2 marks)

....

[Total: 4 marks]

3 Jamie takes out a car loan for £12 500.

His first payment is £440.

After that, he makes 36 equal repayments.

How much is each repayment? (2 marks, ★★★)

....

4 Clara's working contract states:

Working hours: 26 hours per week.

Holiday: 6 weeks per year (including bank holidays).

There are approximately 52 weeks in a year.

How many hours does Clara work each year? (2 marks, ★★★★)

....

Calculating with decimals

Estimate the answer by working out 9×8.
Now you know approximately what your answer should be.

① Work out 9.2×8.3 (2 marks, ★★)

....

② Aisha buys a dress for £22.50 and some shoes for £19.99.

She pays with a £50 note.

How much change should she get? (2 marks, ★★★)

....

WORKIT!

Sam and Jamie visit the funfair. Sam spends three times as much as Jamie. The total amount they spend is £35.60.

How much do they each spend?

Diagrams can help you solve maths problems. This bar shows how much Jamie spends.

J

This bar shows how much Sam spends (because Sam spends three times as much as Jamie).

S	S	S

This bar shows the total amount they spend.

£35.60

J	S	S	S

Jamie spends: $£35.60 \div 4$

£8.90

$$4 \overline{)35.60} = 8.90$$

Sam spends: $£8.90 \times 3$

£26.70

$$890 \times 3 = 2670$$

③ Work out $229.74 \div 6$ (2 marks, ★★★)

....

④ Kirsty and Flo are raising money for charity. Kirsty raises five times as much as Flo. The total amount they raise is £172.50. How much do they each raise? (4 marks, ★★★★★)

....

Rounding and estimation

1. Round 0.7983 to (★★)

   a 3 significant figures (1 mark)

   ....

   b 2 decimal places (1 mark)

   ....

[Total: 2 marks]

2. Estimate an answer to this calculation. (2 marks, ★★★)

$$\frac{9.74 \times 4.02}{7.88}$$

Round each number to the nearest whole number. Then work out the calculation.

....

3. A concert ticket costs £39.50.

   489 concert tickets are sold.

   The concert organisers have costs of £12 500. (★★★★★)

   a Estimate the amount of profit the organisers make. (3 marks)

   ....

   b Is your answer to part **a** an underestimate or an overestimate? (1 mark)

   Give a reason for your answer.

   ....

   ....

[Total: 4 marks]

SNAP IT! Underestimate or overestimate

Profit is the income (amount made) minus any costs.
Underestimate means to estimate something as less than it really is.
Overestimate means to estimate something as more than it really is.

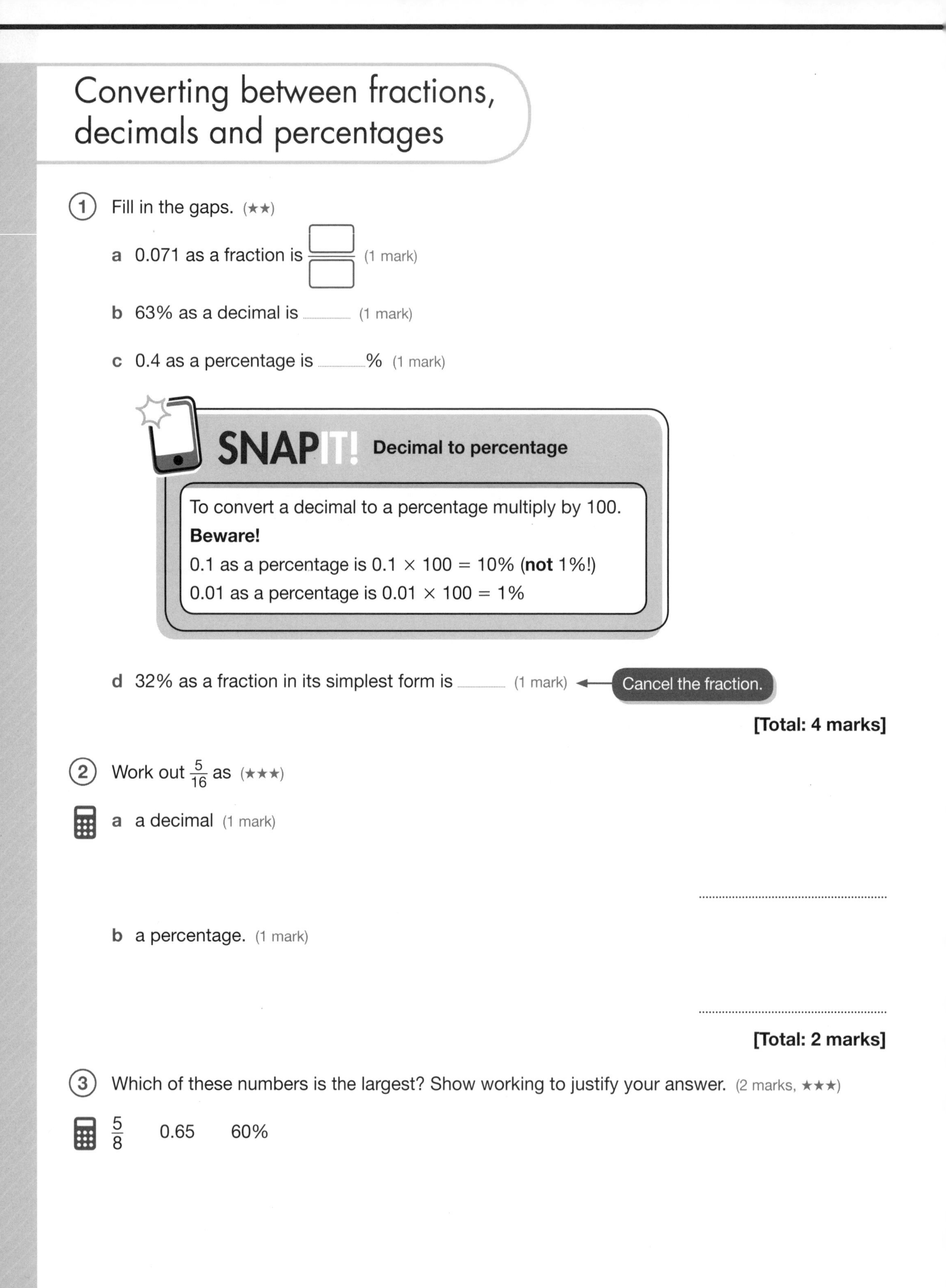

Converting between fractions, decimals and percentages

1. Fill in the gaps. (★★)

   a 0.071 as a fraction is $\frac{\square}{\square}$ (1 mark)

   b 63% as a decimal is .... (1 mark)

   c 0.4 as a percentage is ....% (1 mark)

SNAPIT! Decimal to percentage

To convert a decimal to a percentage multiply by 100.
Beware!
0.1 as a percentage is 0.1 × 100 = 10% (**not** 1%!)
0.01 as a percentage is 0.01 × 100 = 1%

   d 32% as a fraction in its simplest form is .... (1 mark) ← Cancel the fraction.

[Total: 4 marks]

2. Work out $\frac{5}{16}$ as (★★★)

   a a decimal (1 mark)

   ....

   b a percentage. (1 mark)

   ....

[Total: 2 marks]

3. Which of these numbers is the largest? Show working to justify your answer. (2 marks, ★★★)

   $\frac{5}{8}$ 0.65 60%

   ....

Ordering fractions, decimals and percentages

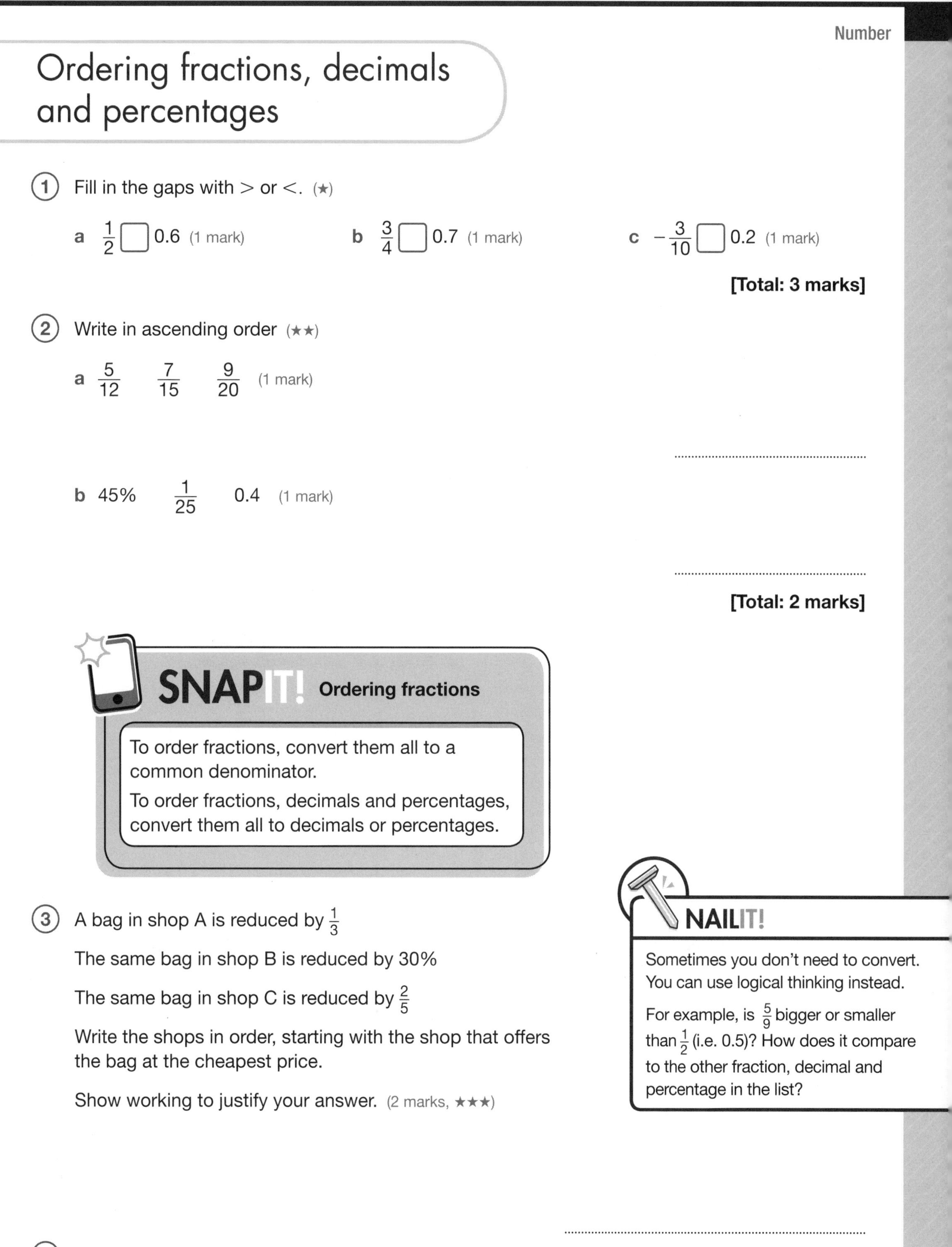

1. Fill in the gaps with > or <. (★)

   a $\frac{1}{2}$ ☐ 0.6 (1 mark)

   b $\frac{3}{4}$ ☐ 0.7 (1 mark)

   c $-\frac{3}{10}$ ☐ 0.2 (1 mark)

[Total: 3 marks]

2. Write in ascending order (★★)

   a $\frac{5}{12}$ $\frac{7}{15}$ $\frac{9}{20}$ (1 mark)

   ....

   b 45% $\frac{1}{25}$ 0.4 (1 mark)

   ....

[Total: 2 marks]

SNAP IT! Ordering fractions

To order fractions, convert them all to a common denominator.

To order fractions, decimals and percentages, convert them all to decimals or percentages.

3. A bag in shop A is reduced by $\frac{1}{3}$

   The same bag in shop B is reduced by 30%

   The same bag in shop C is reduced by $\frac{2}{5}$

   Write the shops in order, starting with the shop that offers the bag at the cheapest price.

   Show working to justify your answer. (2 marks, ★★★)

   ....

NAIL IT!

Sometimes you don't need to convert. You can use logical thinking instead.

For example, is $\frac{5}{9}$ bigger or smaller than $\frac{1}{2}$ (i.e. 0.5)? How does it compare to the other fraction, decimal and percentage in the list?

4. Write in descending order (2 marks, ★★★★)

   0.38 $\frac{5}{9}$ 38.5% $\frac{3}{10}$

   ....

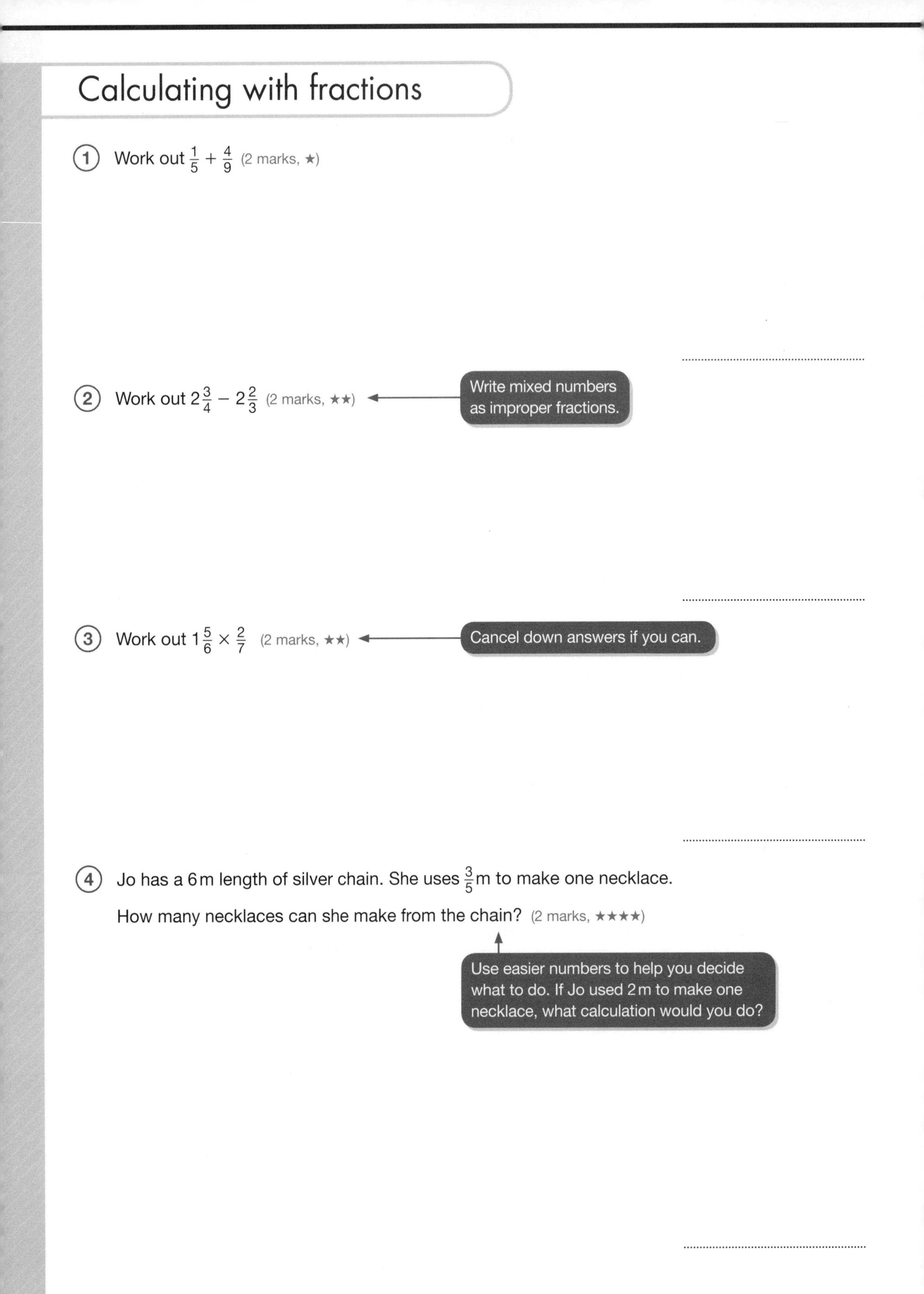

Calculating with fractions

1. Work out $\frac{1}{5} + \frac{4}{9}$ (2 marks, ★)

....

2. Work out $2\frac{3}{4} - 2\frac{2}{3}$ (2 marks, ★★)

Write mixed numbers as improper fractions.

....

3. Work out $1\frac{5}{6} \times \frac{2}{7}$ (2 marks, ★★)

Cancel down answers if you can.

....

4. Jo has a 6 m length of silver chain. She uses $\frac{3}{5}$ m to make one necklace.

How many necklaces can she make from the chain? (2 marks, ★★★★)

Use easier numbers to help you decide what to do. If Jo used 2 m to make one necklace, what calculation would you do?

....

Percentages

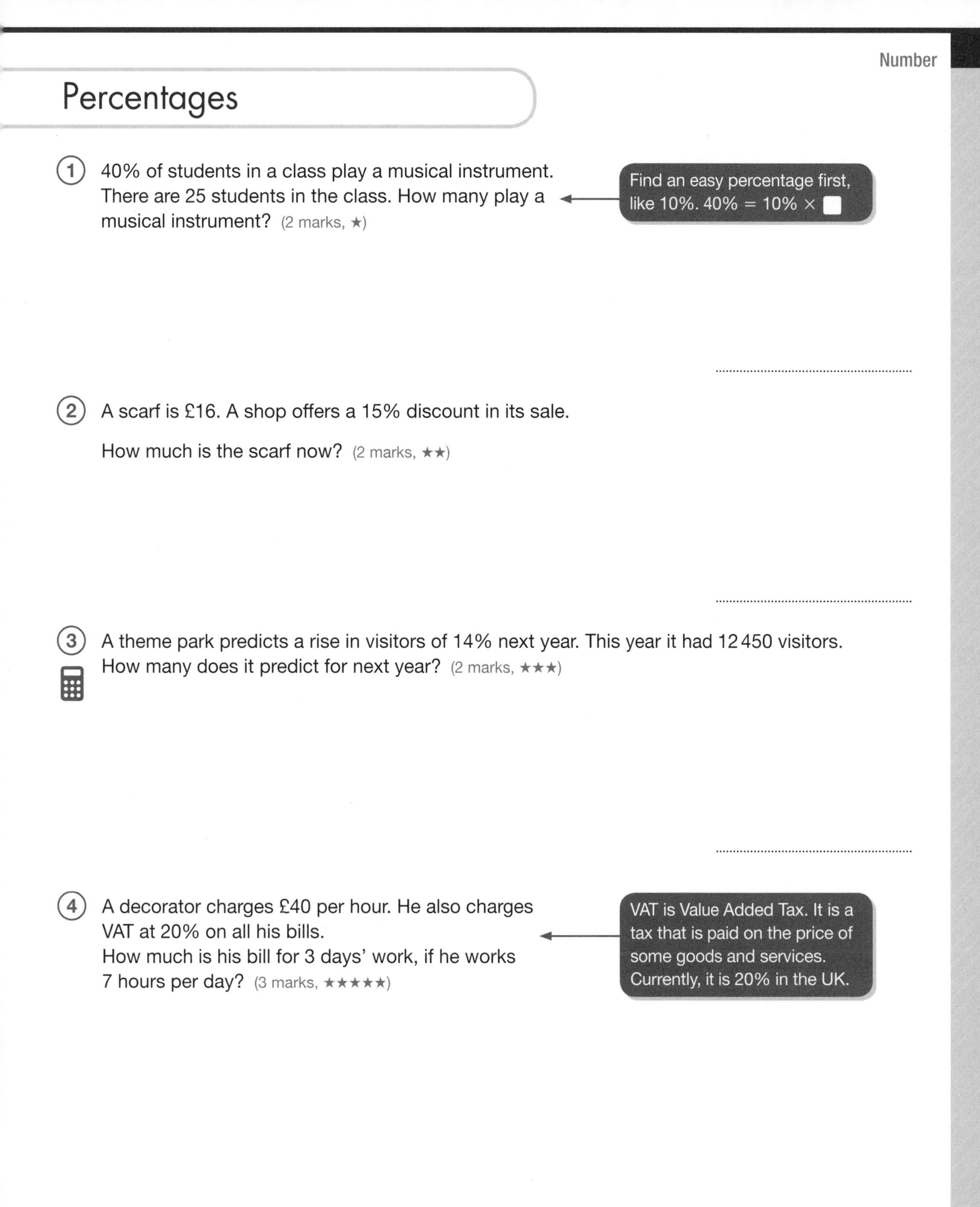

1. 40% of students in a class play a musical instrument. There are 25 students in the class. How many play a musical instrument? (2 marks, ★)

   Find an easy percentage first, like 10%. 40% = 10% × ☐

   ....

2. A scarf is £16. A shop offers a 15% discount in its sale.

   How much is the scarf now? (2 marks, ★★)

   ....

3. A theme park predicts a rise in visitors of 14% next year. This year it had 12 450 visitors. How many does it predict for next year? (2 marks, ★★★)

   ....

4. A decorator charges £40 per hour. He also charges VAT at 20% on all his bills.
   How much is his bill for 3 days' work, if he works 7 hours per day? (3 marks, ★★★★★)

   VAT is Value Added Tax. It is a tax that is paid on the price of some goods and services. Currently, it is 20% in the UK.

   ....

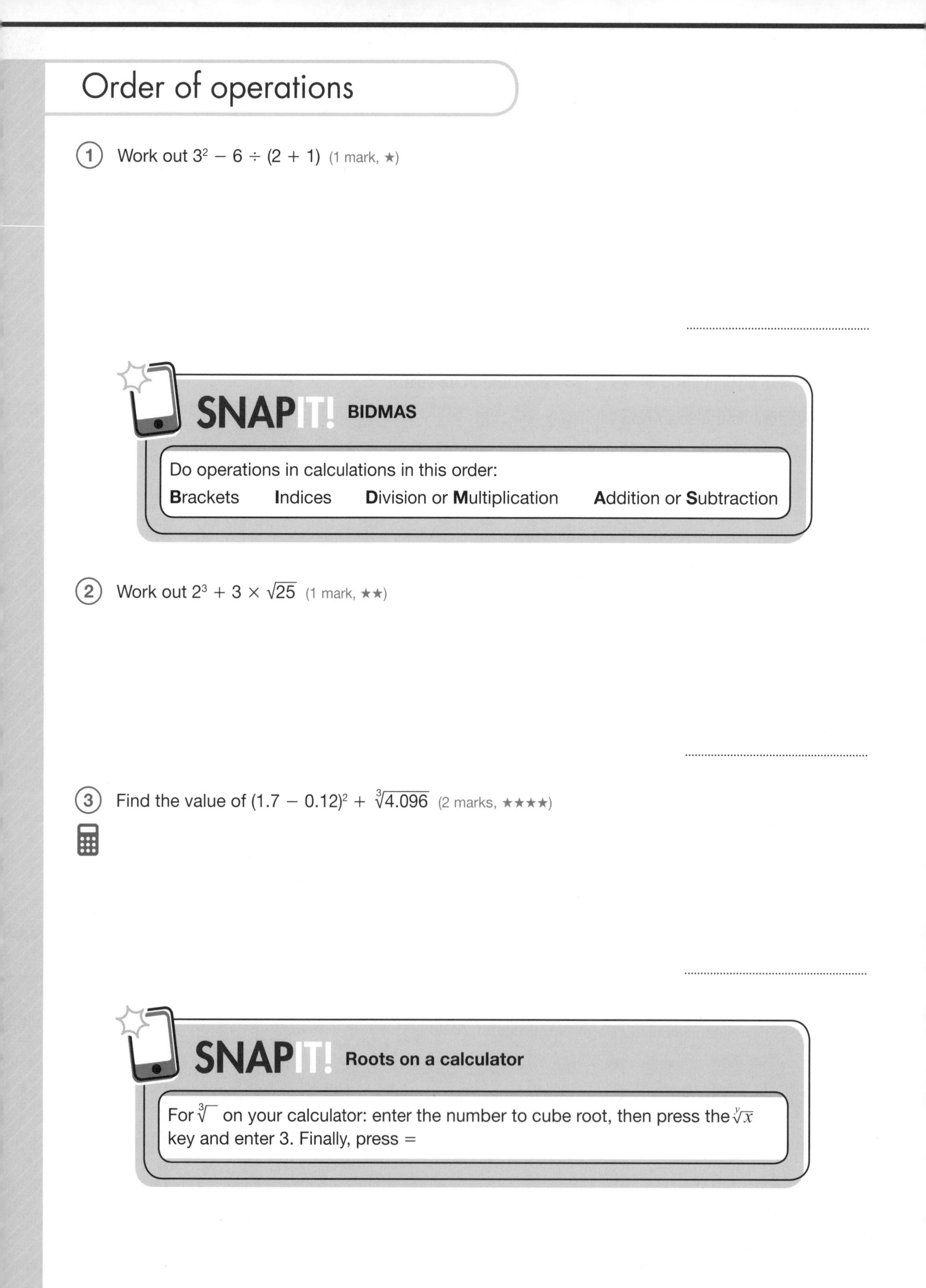

Order of operations

(1) Work out $3^2 - 6 \div (2 + 1)$ (1 mark, ★)

....

SNAPIT! BIDMAS

Do operations in calculations in this order:

Brackets **I**ndices **D**ivision or **M**ultiplication **A**ddition or **S**ubtraction

(2) Work out $2^3 + 3 \times \sqrt{25}$ (1 mark, ★★)

....

(3) Find the value of $(1.7 - 0.12)^2 + \sqrt[3]{4.096}$ (2 marks, ★★★★)

....

SNAPIT! Roots on a calculator

For $\sqrt[3]{\ }$ on your calculator: enter the number to cube root, then press the $\sqrt[y]{x}$ key and enter 3. Finally, press =

Exact solutions

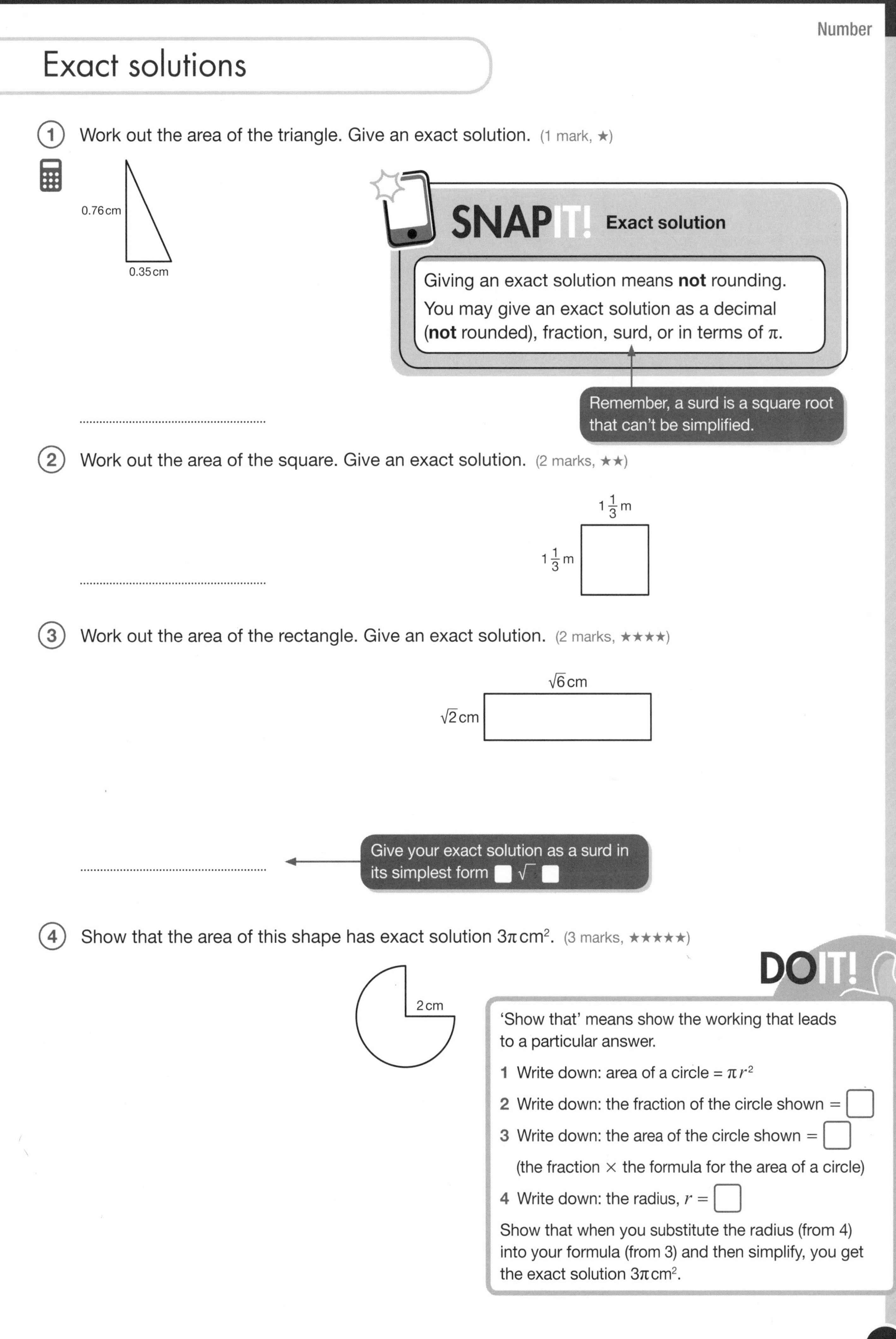

① Work out the area of the triangle. Give an exact solution. (1 mark, ★)

SNAP IT! **Exact solution**

Giving an exact solution means **not** rounding.
You may give an exact solution as a decimal (**not** rounded), fraction, surd, or in terms of π.

Remember, a surd is a square root that can't be simplified.

....

② Work out the area of the square. Give an exact solution. (2 marks, ★★)

....

③ Work out the area of the rectangle. Give an exact solution. (2 marks, ★★★★)

.... ← Give your exact solution as a surd in its simplest form $\square\sqrt{\square}$

④ Show that the area of this shape has exact solution $3\pi\,\text{cm}^2$. (3 marks, ★★★★★)

DO IT!

'Show that' means show the working that leads to a particular answer.

1 Write down: area of a circle = πr^2
2 Write down: the fraction of the circle shown = ☐
3 Write down: the area of the circle shown = ☐
(the fraction × the formula for the area of a circle)
4 Write down: the radius, r = ☐

Show that when you substitute the radius (from 4) into your formula (from 3) and then simplify, you get the exact solution $3\pi\,\text{cm}^2$.

Indices and roots

1. Write these in index form. (★★)

   a $7 \times 7 \times 7 \times 7$ (1 mark)

   ....

   b $\frac{1}{5 \times 5 \times 5}$ (1 mark)

   ....

[Total: 2 marks]

2. Write down the value of (★★★)

   a 2^4 (1 mark)

   ....

   b 10^{-2} (2 marks)

   ....

[Total: 3 marks]

3. Write the following in order of size, starting with the smallest.

   2^3 3^{-2} $\sqrt{25}$ $\sqrt[3]{27}$

   Show working to justify your answer. (3 marks, ★★★★)

   ....

4. Simplify $\frac{9^5}{9^3 \times 9^2}$ (2 marks, ★★★★)

   ....

Standard form

1. Write 2.75×10^3 as an ordinary number. (1 mark, ★)

   ....

2. The average distance from the Sun to the Earth is 150 000 000 km.

   Write this number in standard form. (2 marks, ★★★)

   ....

NAIL IT!

For standard form, write the digit as a number between 1 and 10, and then multiply by a power of 10.

3. Write 0.00642 in standard form. (2 marks, ★★★)

   ....

4. A snail travels 1.4×10^{-5} km in 1 minute. How far does it travel in 20 minutes?

   Write your answer in standard form. (2 marks, ★★★★★)

   ....

Listing strategies

1. Write down all the ways these cards can be arranged. (2 marks, ★★)

   2 5 9

   ....

   ....

   ....

When listing, work logically. For this question
- list all the ways the cards can be arranged when the first card is 2
- now list all the ways the cards can be arranged when the first card is 5
- now list all the ways the cards can be arranged when the first card is 9.

2. Pip spins these 3-sided and 4-sided spinners, then adds the scores. (★★★)

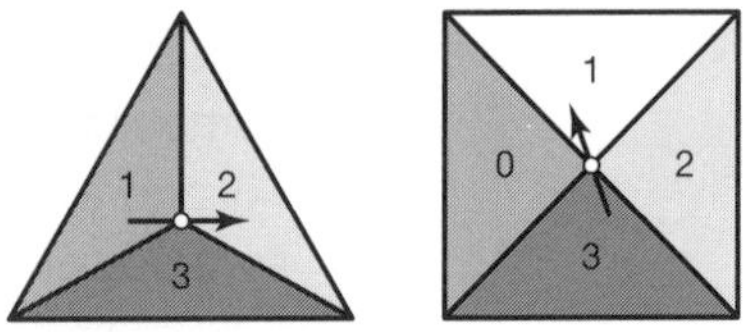

   a Complete the sample space to show all possible outcomes. (2 marks)

		4-sided spinner			
		0	1	2	3
3-sided spinner	1	1	2		
	2	2			
	3				

   b How many different ways can Pip score 3? (1 mark)

   ....

[Total: 3 marks]

3. Draw a sample space for spinning a coin and throwing a dice. (3 marks, ★★★★)

4. A children's menu offers

   Starter – soup (s) or bread (b)

   Main – pasta (p) or fish fingers (f)

   Dessert – jelly (j) or ice-cream (i)

   List all the different options. Use the initials given in brackets. (3 marks, ★★★★)

   ....

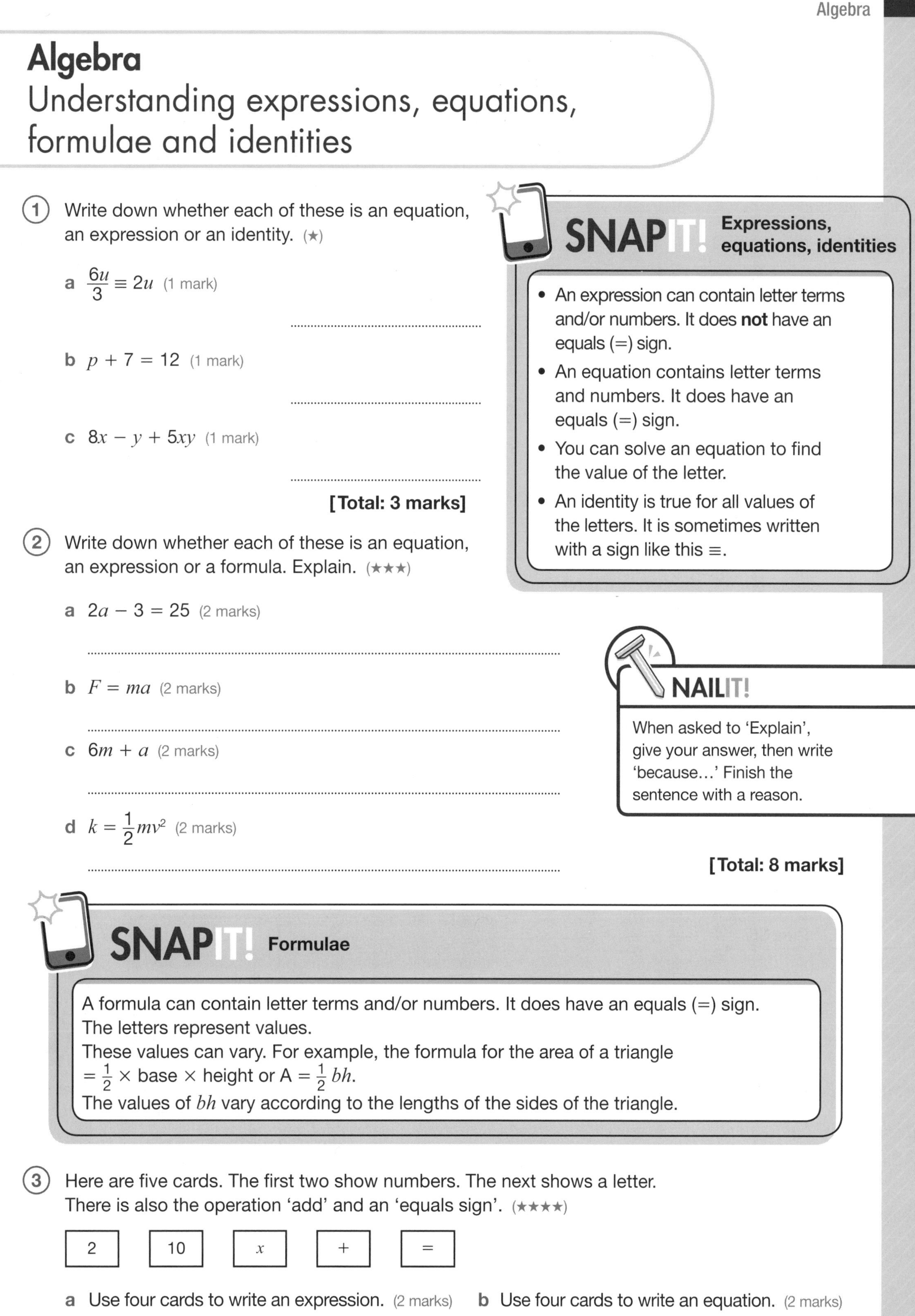

Algebra
Understanding expressions, equations, formulae and identities

1. Write down whether each of these is an equation, an expression or an identity. (★)

   a $\frac{6u}{3} \equiv 2u$ (1 mark)

   ....

   b $p + 7 = 12$ (1 mark)

   ....

   c $8x - y + 5xy$ (1 mark)

   ....

   [Total: 3 marks]

SNAPIT! Expressions, equations, identities

- An expression can contain letter terms and/or numbers. It does **not** have an equals (=) sign.
- An equation contains letter terms and numbers. It does have an equals (=) sign.
- You can solve an equation to find the value of the letter.
- An identity is true for all values of the letters. It is sometimes written with a sign like this ≡.

2. Write down whether each of these is an equation, an expression or a formula. Explain. (★★★)

   a $2a - 3 = 25$ (2 marks)

   ....

   b $F = ma$ (2 marks)

   ....

   c $6m + a$ (2 marks)

   ....

   d $k = \frac{1}{2}mv^2$ (2 marks)

   ....

   [Total: 8 marks]

NAILIT!

When asked to 'Explain', give your answer, then write 'because…' Finish the sentence with a reason.

SNAPIT! Formulae

A formula can contain letter terms and/or numbers. It does have an equals (=) sign.
The letters represent values.
These values can vary. For example, the formula for the area of a triangle $= \frac{1}{2} \times$ base $\times$ height or $A = \frac{1}{2}bh$.
The values of bh vary according to the lengths of the sides of the triangle.

3. Here are five cards. The first two show numbers. The next shows a letter. There is also the operation 'add' and an 'equals sign'. (★★★★)

2	10	x	+	=

   a Use four cards to write an expression. (2 marks)

   ....

   b Use four cards to write an equation. (2 marks)

   ....

   [Total: 4 marks]

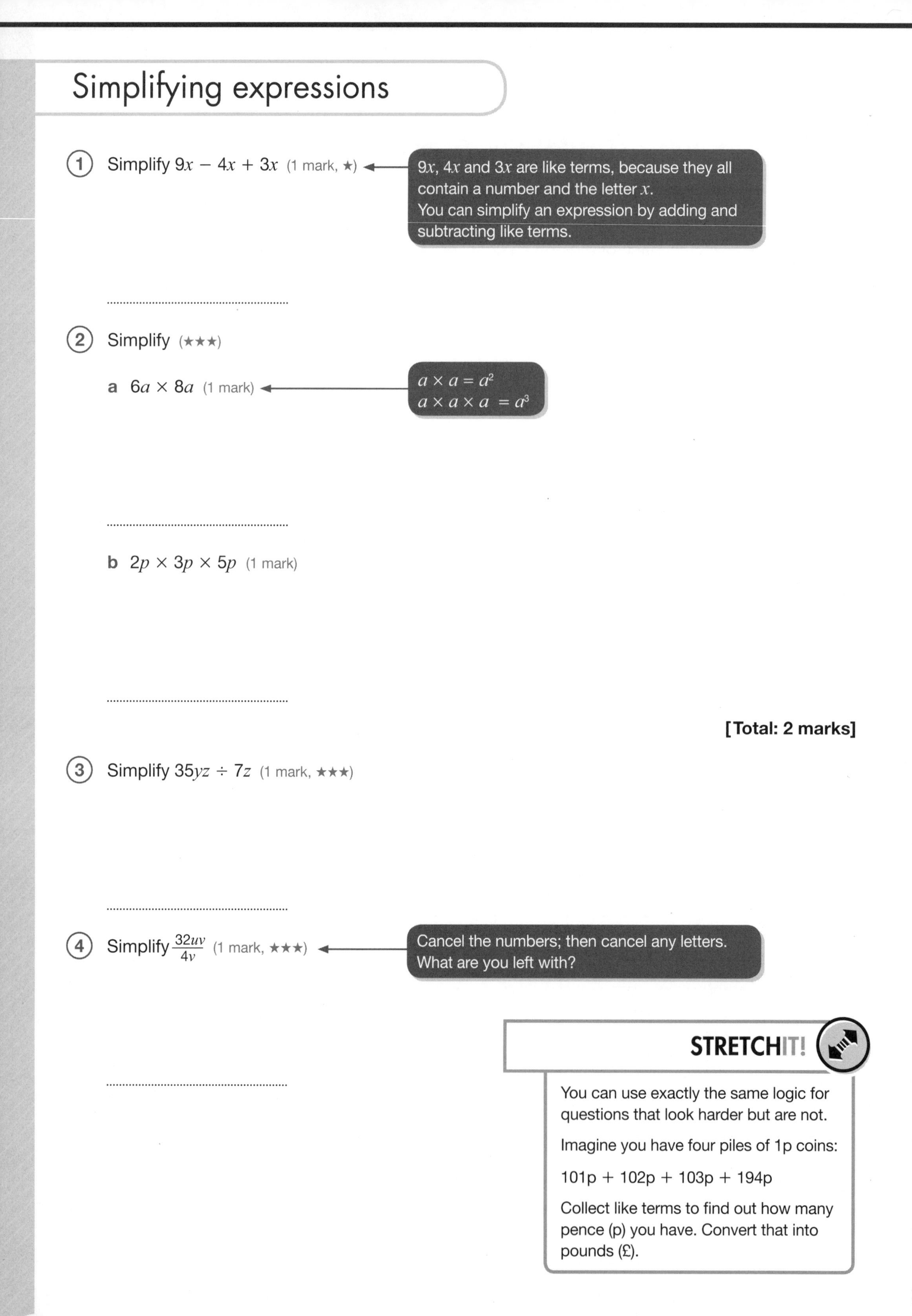

Simplifying expressions

(1) Simplify $9x - 4x + 3x$ (1 mark, ★)

$9x$, $4x$ and $3x$ are like terms, because they all contain a number and the letter x.
You can simplify an expression by adding and subtracting like terms.

....

(2) Simplify (★★★)

a $6a \times 8a$ (1 mark)

$a \times a = a^2$
$a \times a \times a = a^3$

....

b $2p \times 3p \times 5p$ (1 mark)

....

[Total: 2 marks]

(3) Simplify $35yz \div 7z$ (1 mark, ★★★)

....

(4) Simplify $\frac{32uv}{4v}$ (1 mark, ★★★)

Cancel the numbers; then cancel any letters. What are you left with?

....

STRETCH IT!

You can use exactly the same logic for questions that look harder but are not.

Imagine you have four piles of 1p coins:

101p + 102p + 103p + 194p

Collect like terms to find out how many pence (p) you have. Convert that into pounds (£).

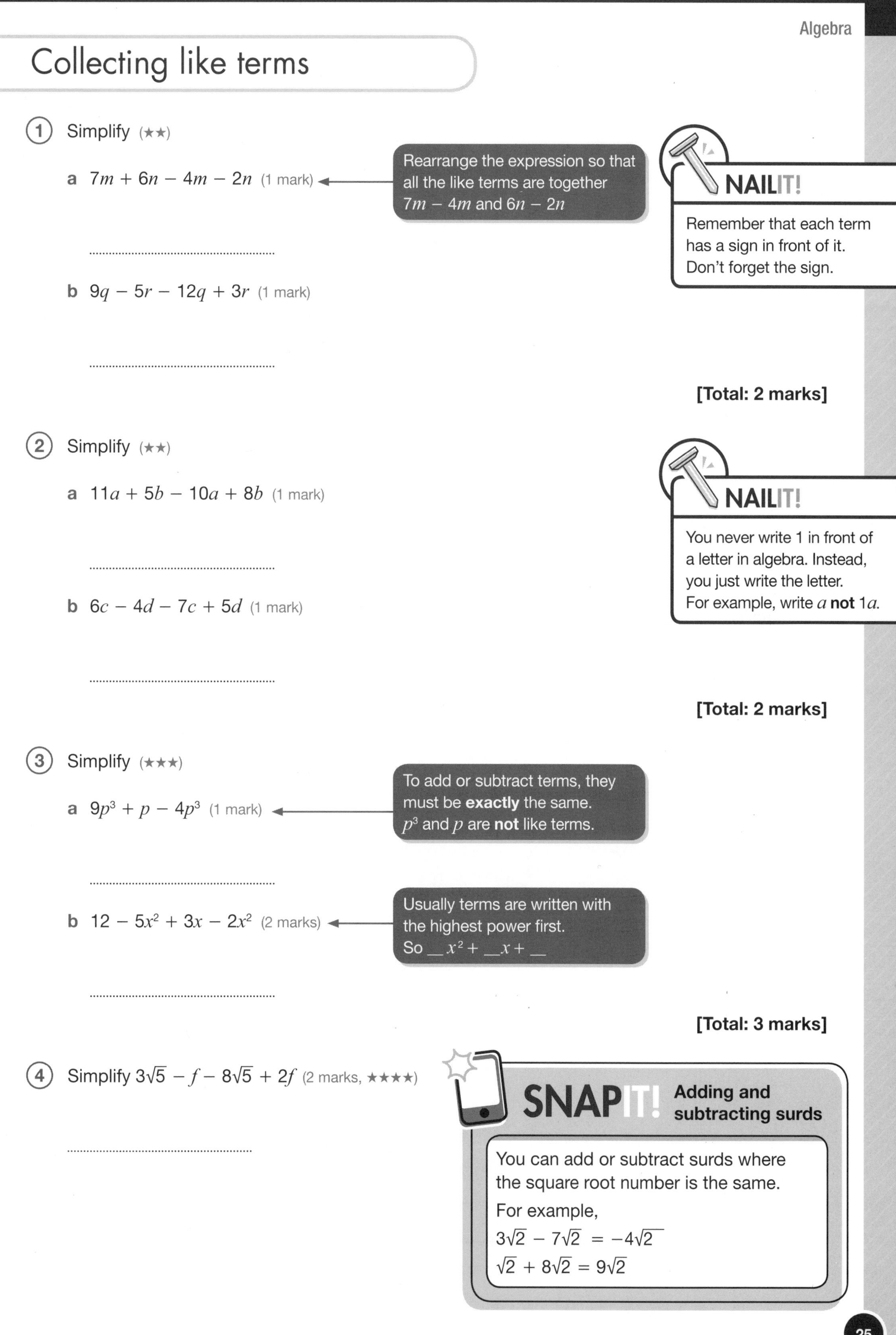

Collecting like terms

1 Simplify (★★)

a $7m + 6n - 4m - 2n$ (1 mark)

Rearrange the expression so that all the like terms are together $7m - 4m$ and $6n - 2n$

....

b $9q - 5r - 12q + 3r$ (1 mark)

....

NAIL IT!

Remember that each term has a sign in front of it. Don't forget the sign.

[Total: 2 marks]

2 Simplify (★★)

a $11a + 5b - 10a + 8b$ (1 mark)

....

b $6c - 4d - 7c + 5d$ (1 mark)

....

NAIL IT!

You never write 1 in front of a letter in algebra. Instead, you just write the letter. For example, write a **not** $1a$.

[Total: 2 marks]

3 Simplify (★★★)

a $9p^3 + p - 4p^3$ (1 mark)

To add or subtract terms, they must be **exactly** the same. p^3 and p are **not** like terms.

....

b $12 - 5x^2 + 3x - 2x^2$ (2 marks)

Usually terms are written with the highest power first. So $\_\_x^2 + \_\_x + \_\_$

....

[Total: 3 marks]

4 Simplify $3\sqrt{5} - f - 8\sqrt{5} + 2f$ (2 marks, ★★★★)

....

SNAP IT! Adding and subtracting surds

You can add or subtract surds where the square root number is the same.

For example,

$3\sqrt{2} - 7\sqrt{2} = -4\sqrt{2}$

$\sqrt{2} + 8\sqrt{2} = 9\sqrt{2}$

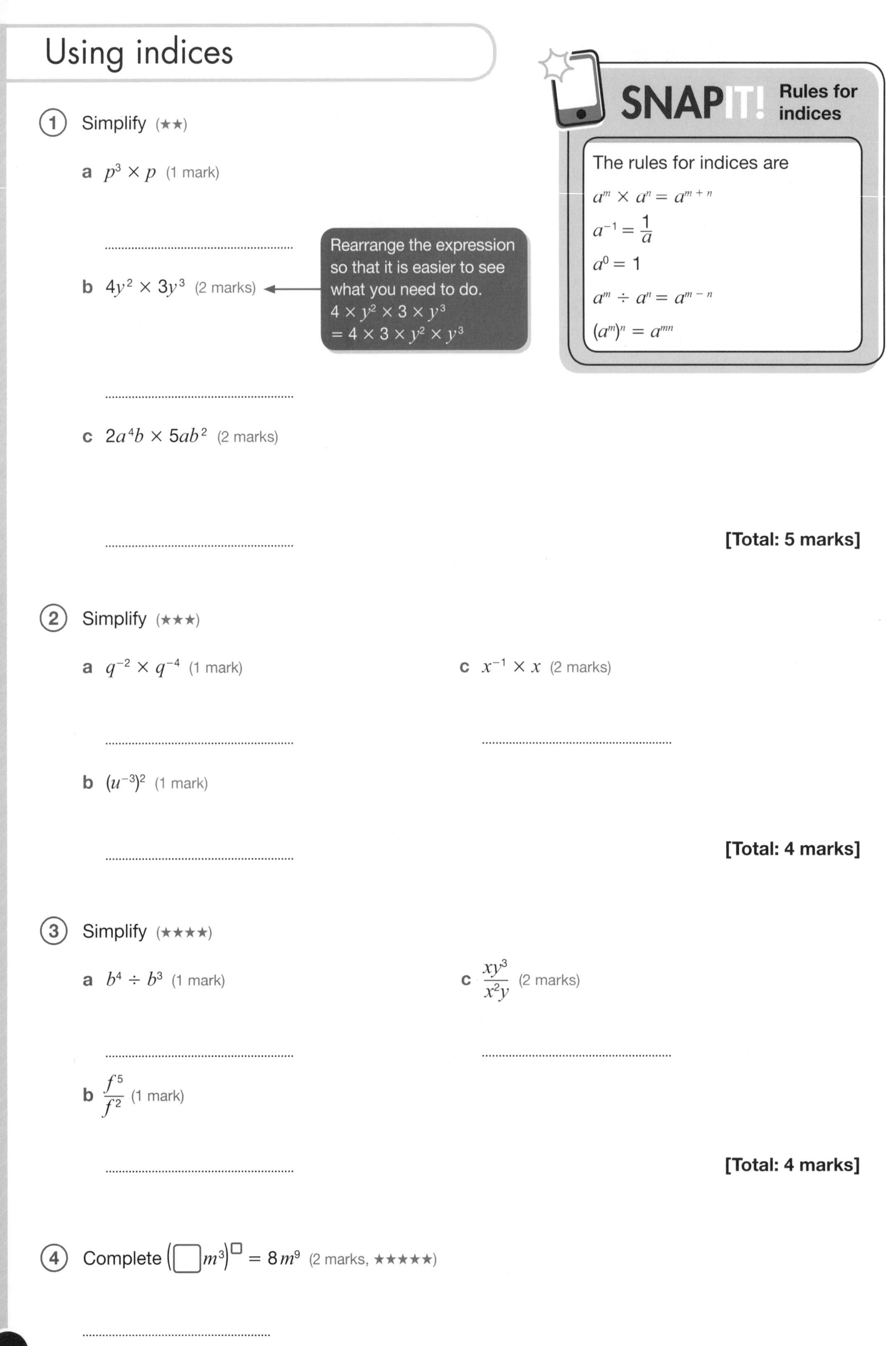

Using indices

SNAPIT! **Rules for indices**

The rules for indices are

$a^m \times a^n = a^{m+n}$

$a^{-1} = \frac{1}{a}$

$a^0 = 1$

$a^m \div a^n = a^{m-n}$

$(a^m)^n = a^{mn}$

1 Simplify (★★)

a $p^3 \times p$ (1 mark)

b $4y^2 \times 3y^3$ (2 marks)

Rearrange the expression so that it is easier to see what you need to do.
$4 \times y^2 \times 3 \times y^3$
$= 4 \times 3 \times y^2 \times y^3$

c $2a^4b \times 5ab^2$ (2 marks)

[Total: 5 marks]

2 Simplify (★★★)

a $q^{-2} \times q^{-4}$ (1 mark)

b $(u^{-3})^2$ (1 mark)

c $x^{-1} \times x$ (2 marks)

[Total: 4 marks]

3 Simplify (★★★★)

a $b^4 \div b^3$ (1 mark)

b $\frac{f^5}{f^2}$ (1 mark)

c $\frac{xy^3}{x^2y}$ (2 marks)

[Total: 4 marks]

4 Complete $(\square m^3)^{\square} = 8m^9$ (2 marks, ★★★★★)

Expanding brackets

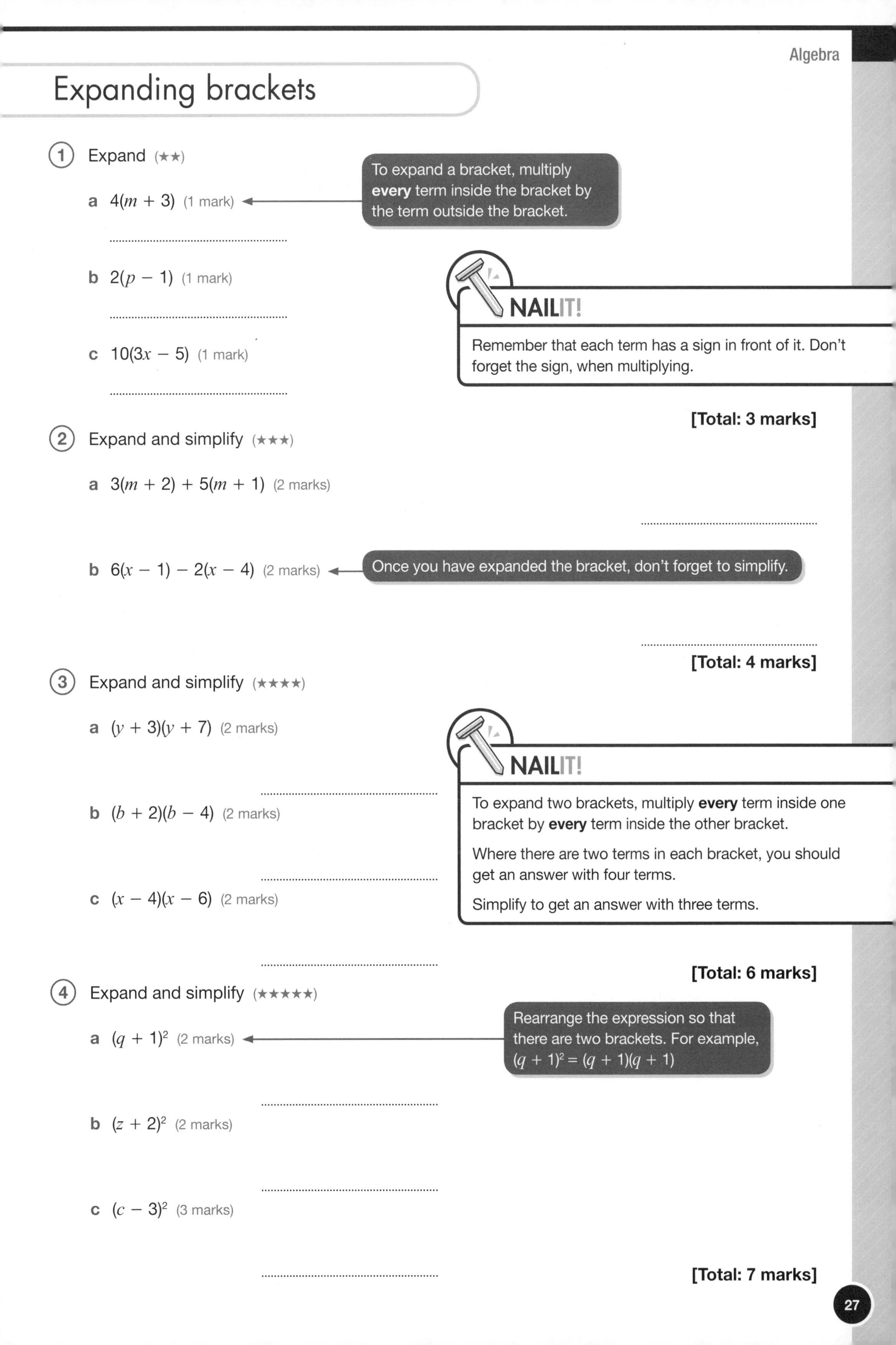

1. Expand (★★)

To expand a bracket, multiply **every** term inside the bracket by the term outside the bracket.

a $4(m + 3)$ (1 mark)

....

b $2(p - 1)$ (1 mark)

....

c $10(3x - 5)$ (1 mark)

....

NAILIT!

Remember that each term has a sign in front of it. Don't forget the sign, when multiplying.

[Total: 3 marks]

2. Expand and simplify (★★★)

a $3(m + 2) + 5(m + 1)$ (2 marks)

....

b $6(x - 1) - 2(x - 4)$ (2 marks)

Once you have expanded the bracket, don't forget to simplify.

....

[Total: 4 marks]

3. Expand and simplify (★★★★)

a $(y + 3)(y + 7)$ (2 marks)

....

b $(b + 2)(b - 4)$ (2 marks)

....

c $(x - 4)(x - 6)$ (2 marks)

....

NAILIT!

To expand two brackets, multiply **every** term inside one bracket by **every** term inside the other bracket.

Where there are two terms in each bracket, you should get an answer with four terms.

Simplify to get an answer with three terms.

[Total: 6 marks]

4. Expand and simplify (★★★★★)

a $(q + 1)^2$ (2 marks)

Rearrange the expression so that there are two brackets. For example, $(q + 1)^2 = (q + 1)(q + 1)$

....

b $(z + 2)^2$ (2 marks)

....

c $(c - 3)^2$ (3 marks)

....

[Total: 7 marks]

Factorising

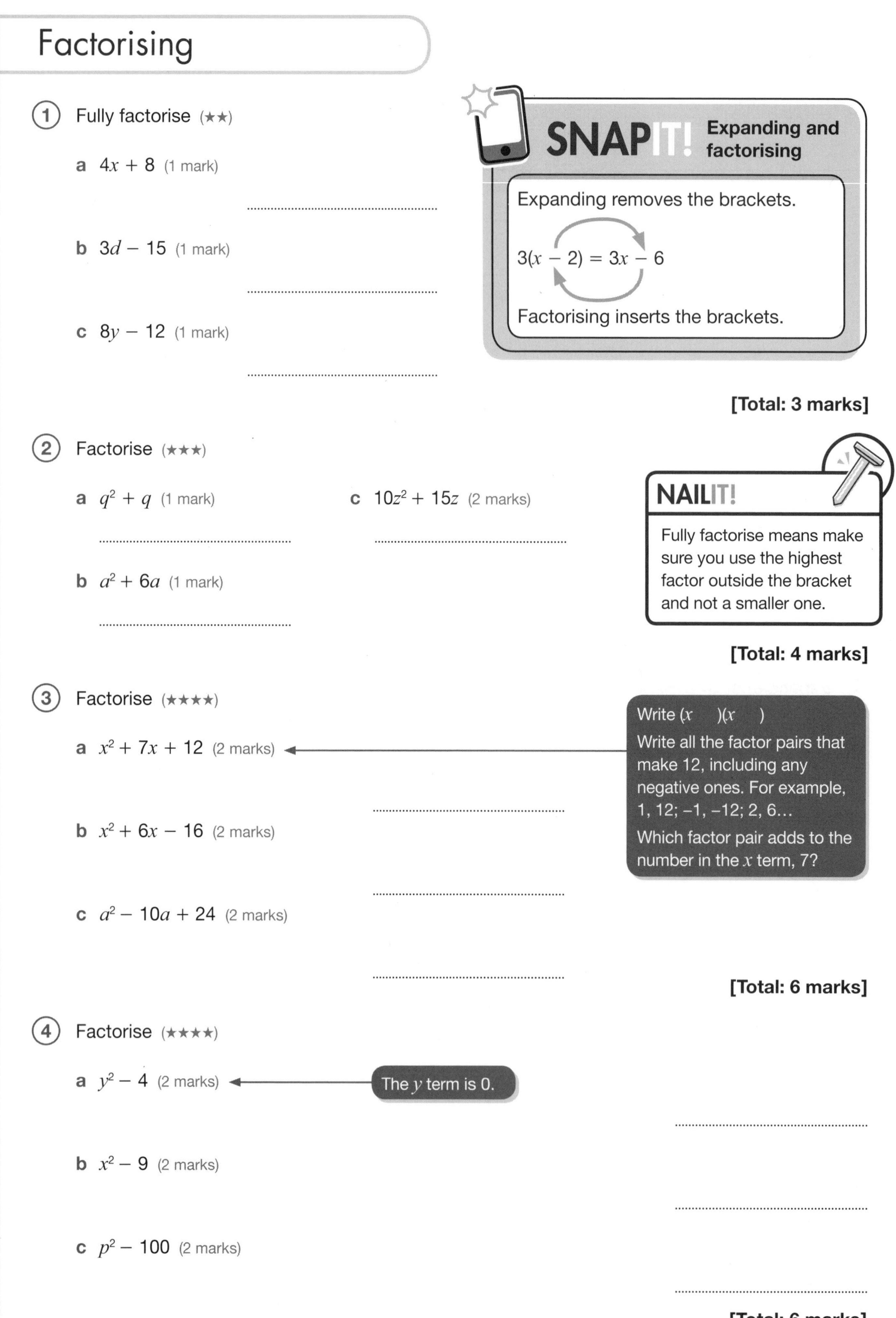

SNAP IT! Expanding and factorising

Expanding removes the brackets.

$3(x - 2) = 3x - 6$

Factorising inserts the brackets.

1 Fully factorise (★★)

a $4x + 8$ (1 mark)

....

b $3d - 15$ (1 mark)

....

c $8y - 12$ (1 mark)

....

[Total: 3 marks]

2 Factorise (★★★)

a $q^2 + q$ (1 mark)

....

b $a^2 + 6a$ (1 mark)

....

c $10z^2 + 15z$ (2 marks)

....

NAIL IT!

Fully factorise means make sure you use the highest factor outside the bracket and not a smaller one.

[Total: 4 marks]

3 Factorise (★★★★)

a $x^2 + 7x + 12$ (2 marks)

Write $(x \quad)(x \quad)$
Write all the factor pairs that make 12, including any negative ones. For example, 1, 12; −1, −12; 2, 6…
Which factor pair adds to the number in the x term, 7?

....

b $x^2 + 6x - 16$ (2 marks)

....

c $a^2 - 10a + 24$ (2 marks)

....

[Total: 6 marks]

4 Factorise (★★★★)

a $y^2 - 4$ (2 marks)

The y term is 0.

....

b $x^2 - 9$ (2 marks)

....

c $p^2 - 100$ (2 marks)

....

[Total: 6 marks]

Substituting into expressions

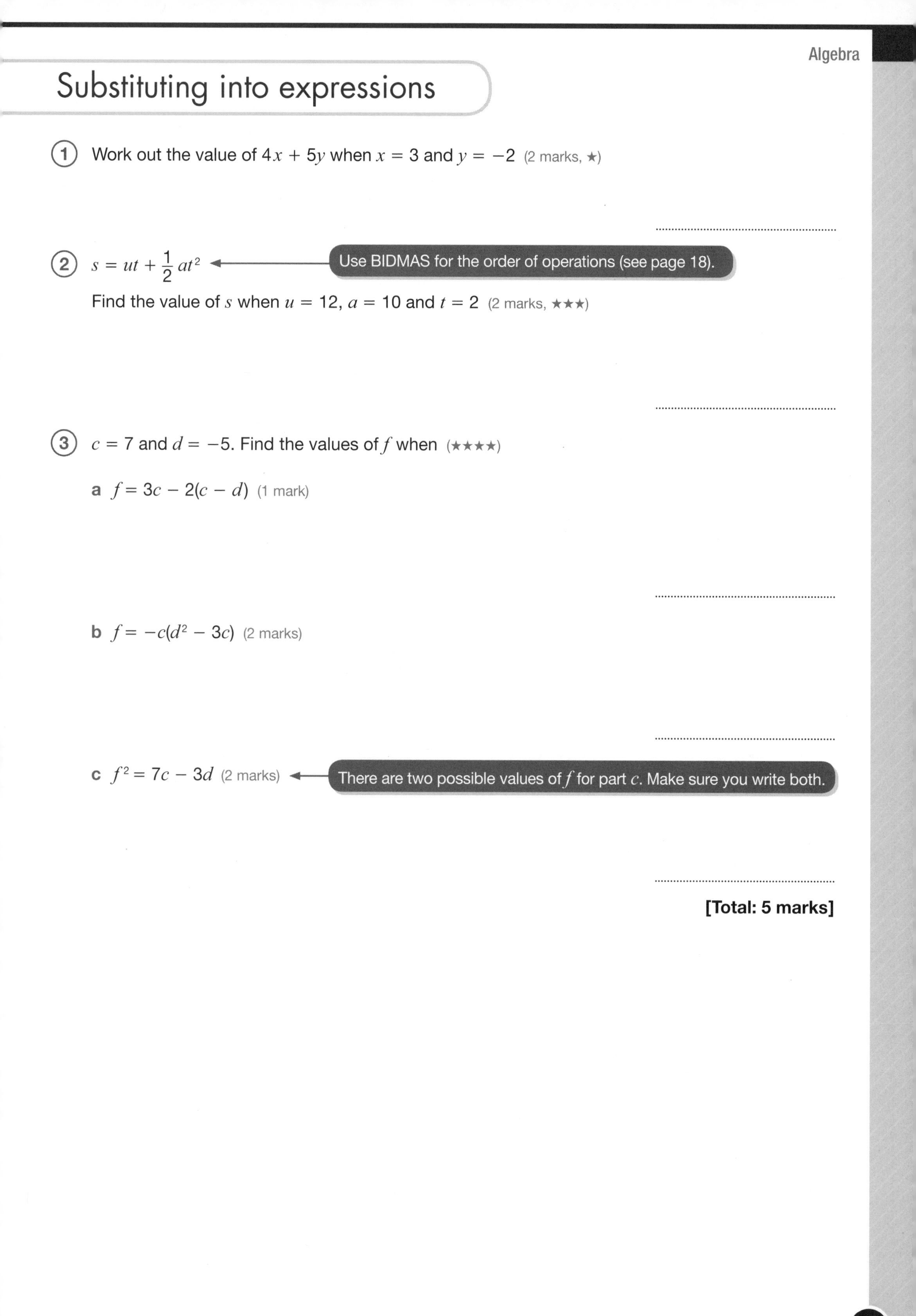

① Work out the value of $4x + 5y$ when $x = 3$ and $y = -2$ (2 marks, ★)

....

② $s = ut + \frac{1}{2}at^2$ ← Use BIDMAS for the order of operations (see page 18).

Find the value of s when $u = 12$, $a = 10$ and $t = 2$ (2 marks, ★★★)

....

③ $c = 7$ and $d = -5$. Find the values of f when (★★★★)

a $f = 3c - 2(c - d)$ (1 mark)

....

b $f = -c(d^2 - 3c)$ (2 marks)

....

c $f^2 = 7c - 3d$ (2 marks) ← There are two possible values of f for part c. Make sure you write both.

....

[Total: 5 marks]

Writing expressions

① a Ali thinks of a number, n. He adds 3 to get a new number.

Write an expression for Ali's new number. (1 mark, ★★)

Pretend Ali's original number is 7. What calculation would you do to work out his new number? Now do the same for the letter n. This is the expression for Ali's new number.

....

b Kamal thinks of a number, n. He doubles it and subtracts 9 to get a new number.

Write an expression for Kamal's new number. (2 marks, ★★)

....

[Total: 3 marks]

② Eve sells hats for £x and scarves for £y. Write an expression for the money she makes when she sells (★★★)

a 1 hat and 1 scarf (1 mark)

....

b 5 hats (1 mark)

....

c 12 hats and 11 scarves. (2 marks)

....

[Total: 4 marks]

③ A parallelogram has side lengths $9p$ and $5p + 2$.

Write an expression for its perimeter.

Give your answer in its simplest form. (4 marks, ★★★★)

Often it can help to draw a diagram. Draw the parallelogram and put all the information on it.

....

④ A rectangular flower bed has side lengths s and $5s + 1$.

Write an expression for its area. (1 mark, ★★★★)

Use a bracket in your expression.

....

Solving linear equations

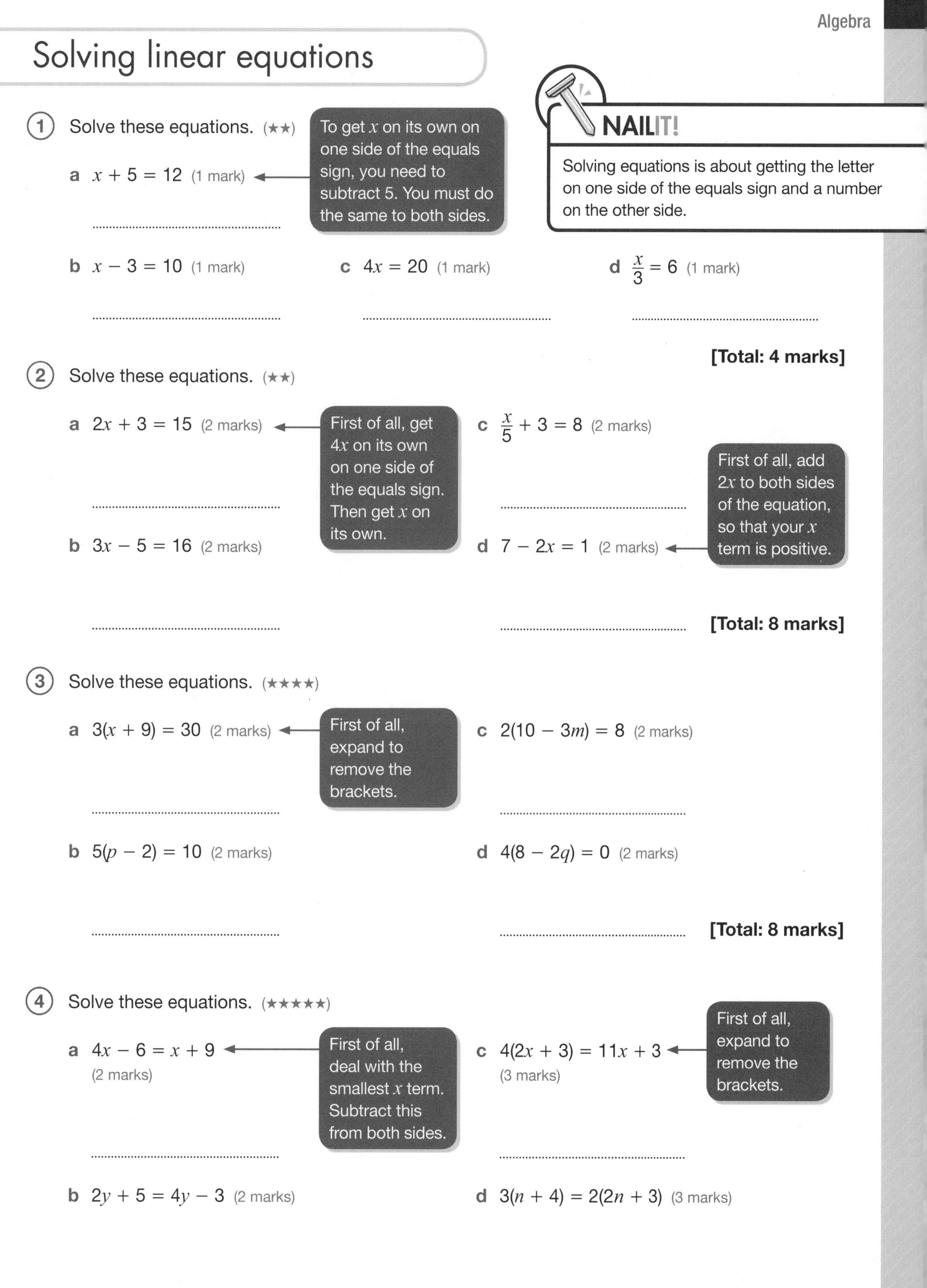

NAIL IT!

Solving equations is about getting the letter on one side of the equals sign and a number on the other side.

1. Solve these equations. (★★)

a $x + 5 = 12$ (1 mark)

To get x on its own on one side of the equals sign, you need to subtract 5. You must do the same to both sides.

....

b $x - 3 = 10$ (1 mark)

....

c $4x = 20$ (1 mark)

....

d $\frac{x}{3} = 6$ (1 mark)

....

[Total: 4 marks]

2. Solve these equations. (★★)

a $2x + 3 = 15$ (2 marks)

First of all, get $4x$ on its own on one side of the equals sign. Then get x on its own.

....

b $3x - 5 = 16$ (2 marks)

....

c $\frac{x}{5} + 3 = 8$ (2 marks)

....

d $7 - 2x = 1$ (2 marks)

First of all, add $2x$ to both sides of the equation, so that your x term is positive.

....

[Total: 8 marks]

3. Solve these equations. (★★★★)

a $3(x + 9) = 30$ (2 marks)

First of all, expand to remove the brackets.

....

b $5(p - 2) = 10$ (2 marks)

....

c $2(10 - 3m) = 8$ (2 marks)

....

d $4(8 - 2q) = 0$ (2 marks)

....

[Total: 8 marks]

4. Solve these equations. (★★★★★)

a $4x - 6 = x + 9$ (2 marks)

First of all, deal with the smallest x term. Subtract this from both sides.

....

b $2y + 5 = 4y - 3$ (2 marks)

....

c $4(2x + 3) = 11x + 3$ (3 marks)

First of all, expand to remove the brackets.

....

d $3(n + 4) = 2(2n + 3)$ (3 marks)

....

[Total: 10 marks]

Writing linear equations

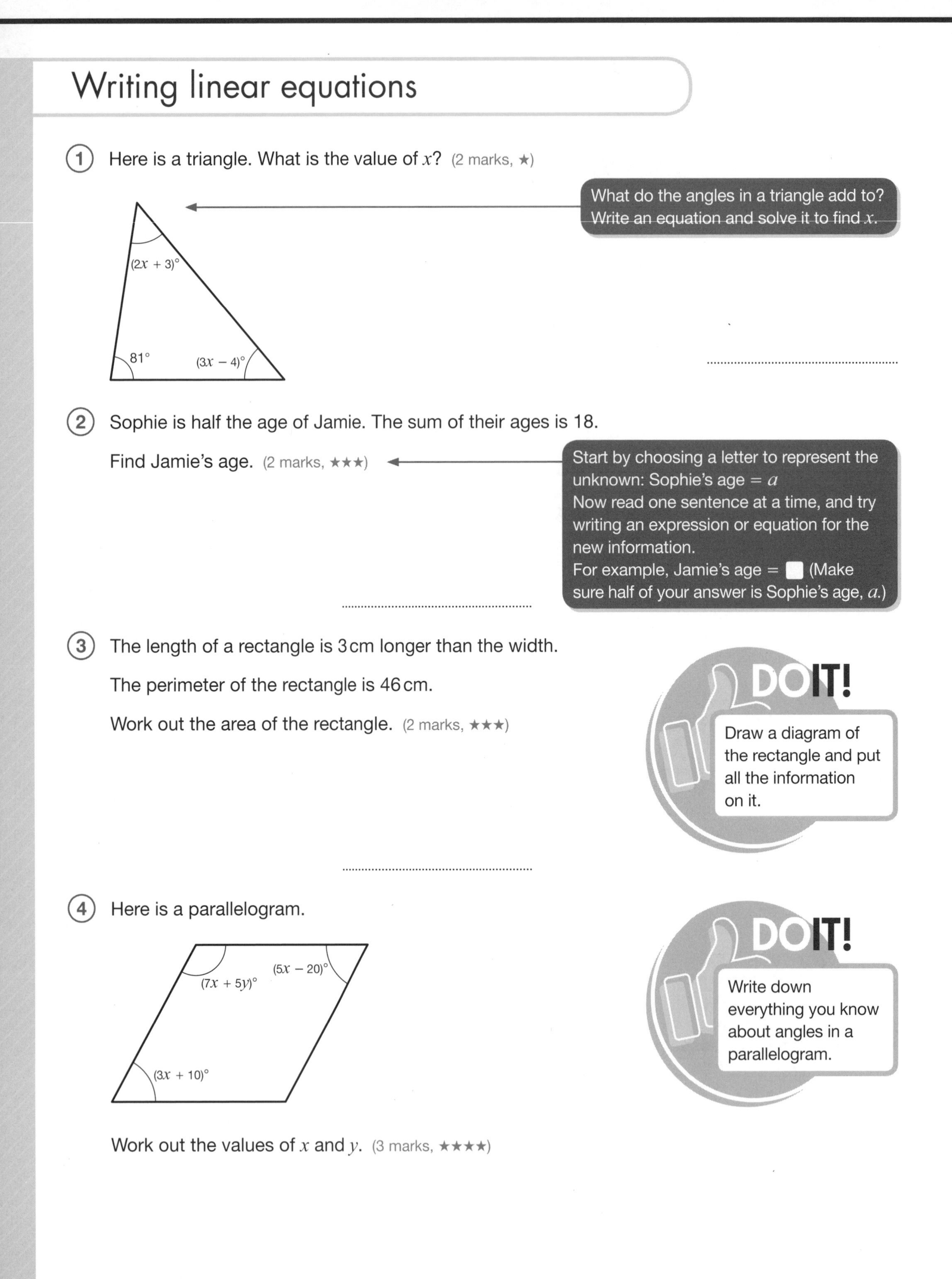

1. Here is a triangle. What is the value of x? (2 marks, ★)

What do the angles in a triangle add to? Write an equation and solve it to find x.

....

2. Sophie is half the age of Jamie. The sum of their ages is 18.

   Find Jamie's age. (2 marks, ★★★)

Start by choosing a letter to represent the unknown: Sophie's age = a
Now read one sentence at a time, and try writing an expression or equation for the new information.
For example, Jamie's age = ▢ (Make sure half of your answer is Sophie's age, a.)

....

3. The length of a rectangle is 3 cm longer than the width.

   The perimeter of the rectangle is 46 cm.

   Work out the area of the rectangle. (2 marks, ★★★)

DO IT! Draw a diagram of the rectangle and put all the information on it.

....

4. Here is a parallelogram.

   Work out the values of x and y. (3 marks, ★★★★)

DO IT! Write down everything you know about angles in a parallelogram.

....

Linear inequalities

SNAPIT! **Integers**

An integer is any whole number.
It can be positive, negative, or zero.

① x is an integer such that $-2 < x \leq 5$ (★★)

a List all the integer values of x. (1 mark)

....

b Show the inequality $-2 < x \leq 5$ on a number line. (2 marks)

[Total: 3 marks]

② Solve each inequality. Show each answer on a number line. (★★★)

a $4x > 20$ (2 marks)

b $3x - 8 \leq 13$ (2 marks)

c $2(x - 3) < 10$ (2 marks)

[Total: 6 marks]

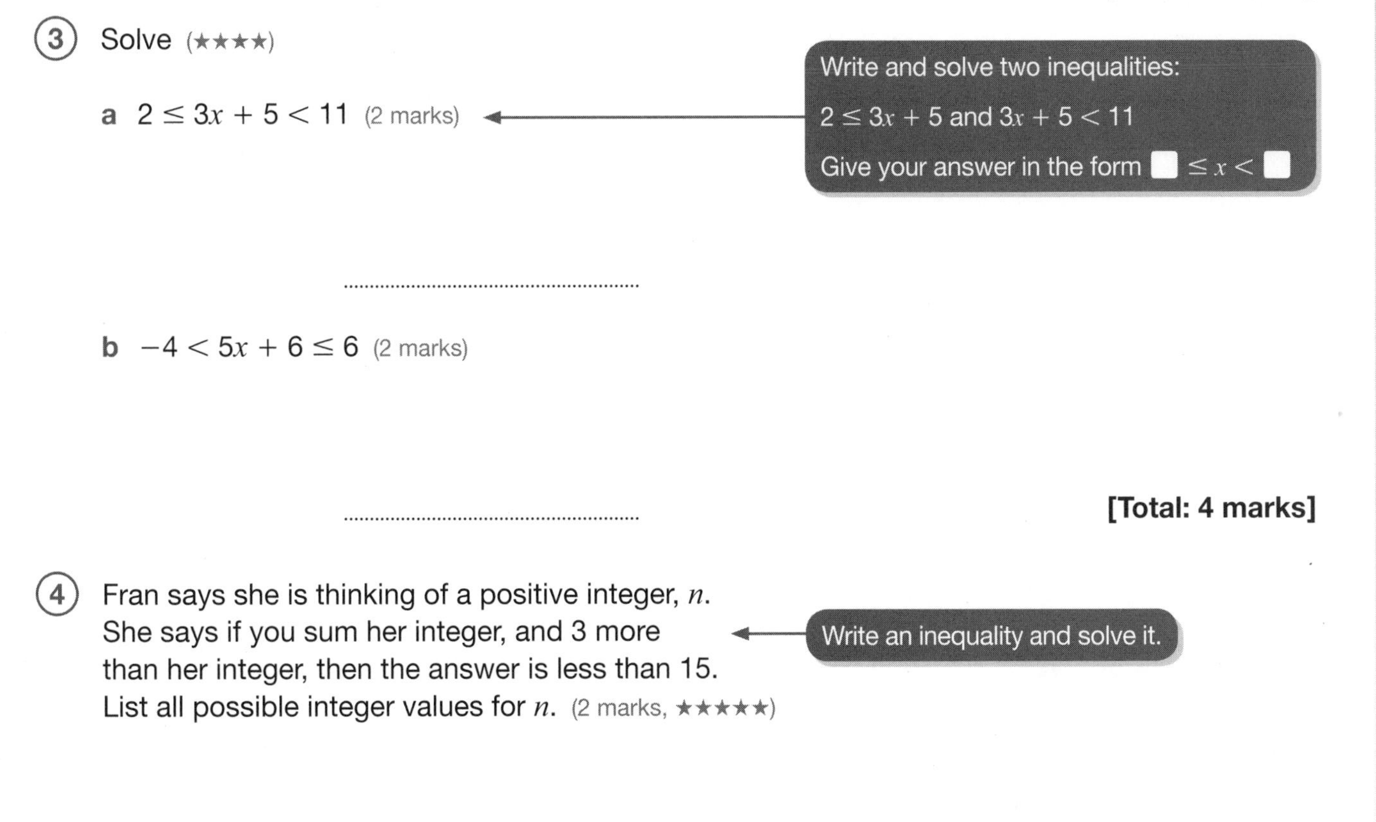

③ Solve (★★★★)

a $2 \leq 3x + 5 < 11$ (2 marks)

> Write and solve two inequalities:
> $2 \leq 3x + 5$ and $3x + 5 < 11$
> Give your answer in the form ☐ $\leq x <$ ☐

....

b $-4 < 5x + 6 \leq 6$ (2 marks)

....

[Total: 4 marks]

④ Fran says she is thinking of a positive integer, n.
She says if you sum her integer, and 3 more than her integer, then the answer is less than 15.
List all possible integer values for n. (2 marks, ★★★★★)

> Write an inequality and solve it.

....

Formulae

① The length of time, in minutes, to roast a chicken is calculated using this formula

$t = 40w + 20$

where t is the time in minutes and w is the weight of the chicken in kg. (★★)

a How long does it take to roast a chicken that weighs 2 kg?

Give your answer in hours and minutes. (2 marks)

....

b What time should a 1.5 kg chicken be put in the oven for a Sunday lunch at 1.30 pm? (2 marks)

....

[Total: 4 marks]

② A locksmith charges £l to change the locks on a door, and then £k for each new key. (★★★)

a Write a formula for C, the cost of using the locksmith, if a customer orders n keys. (3 marks)

....

Use different numbers for l, k and n to help you write a calculation. For example, a locksmith charges £50 to change the locks on a door, and then £5 for each new key. Then substitute the letters back in to write the formula.

b Use your formula to find C, in £, when $l = 90$, $k = 6.5$ and $n = 3$. (2 marks)

....

[Total: 5 marks]

③ Rearrange each formula to make q the subject. (★★★★)

a $p = \frac{qs}{3}$ (2 marks)

....

Decide what is happening to q, and then do the inverse to all terms on both sides of the equation.
What is happening to q:

$\times s \div 3$

So do the inverse:

$\times 3 \div s$

b $p = \frac{q}{r} + t$ (2 marks)

....

c $p = 3(q + r)$ (2 marks)

First of all, multiply out the brackets.

....

d $p = \sqrt{2q}$ (2 marks)

First square both sides.

....

[Total: 8 marks]

SNAPIT! **Subject of a formula**

In a formula the letter on its own on one side of the equals sign is called the subject.

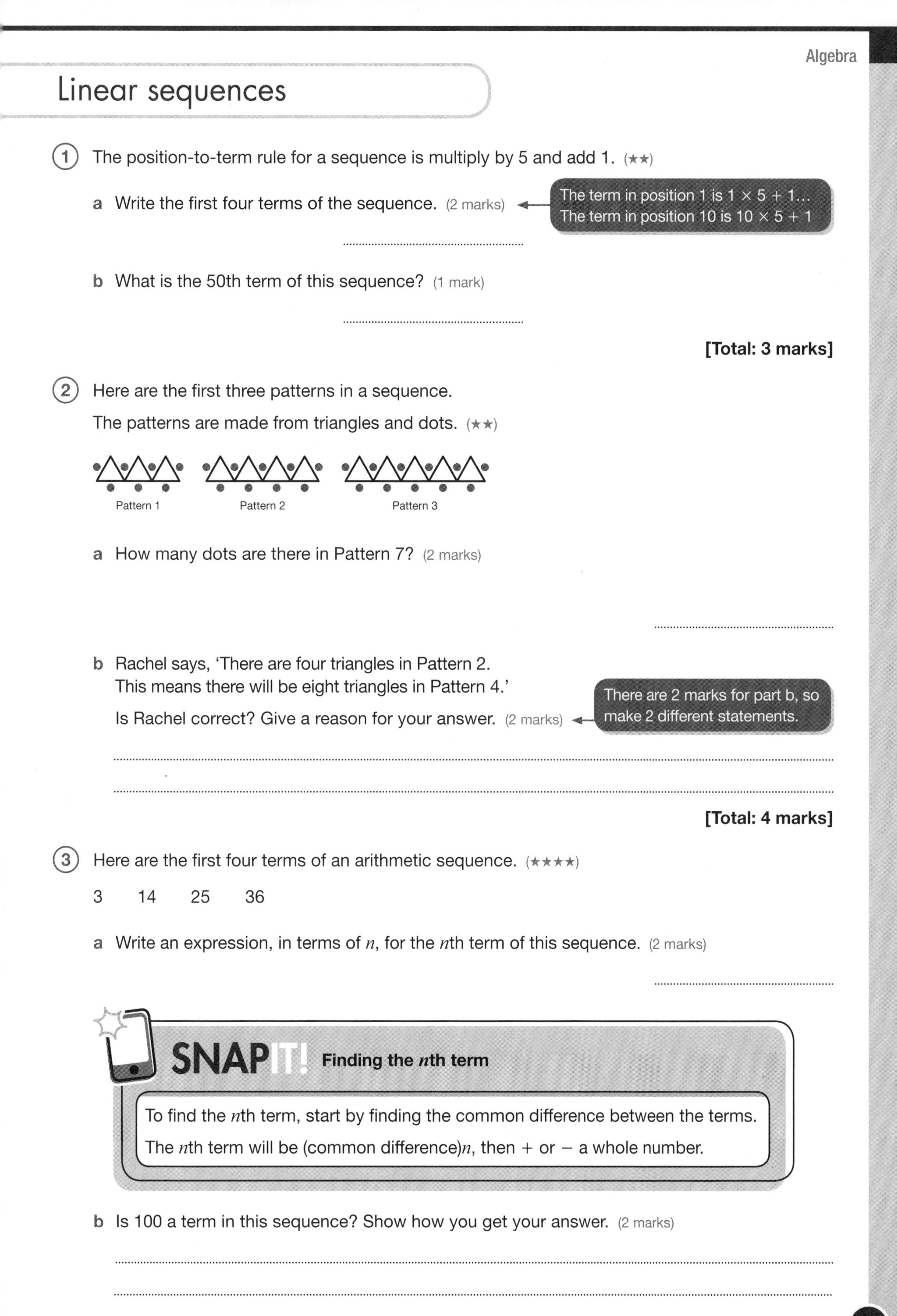

Linear sequences

(1) The position-to-term rule for a sequence is multiply by 5 and add 1. (★★)

a Write the first four terms of the sequence. (2 marks)

The term in position 1 is $1 \times 5 + 1$...
The term in position 10 is $10 \times 5 + 1$

....

b What is the 50th term of this sequence? (1 mark)

....

[Total: 3 marks]

(2) Here are the first three patterns in a sequence.

The patterns are made from triangles and dots. (★★)

a How many dots are there in Pattern 7? (2 marks)

....

b Rachel says, 'There are four triangles in Pattern 2.
This means there will be eight triangles in Pattern 4.'

Is Rachel correct? Give a reason for your answer. (2 marks)

There are 2 marks for part b, so make 2 different statements.

....

....

[Total: 4 marks]

(3) Here are the first four terms of an arithmetic sequence. (★★★★)

3 14 25 36

a Write an expression, in terms of n, for the nth term of this sequence. (2 marks)

....

SNAP IT! **Finding the nth term**

To find the nth term, start by finding the common difference between the terms.

The nth term will be (common difference)n, then + or − a whole number.

b Is 100 a term in this sequence? Show how you get your answer. (2 marks)

....

....

[Total: 4 marks]

Non-linear sequences

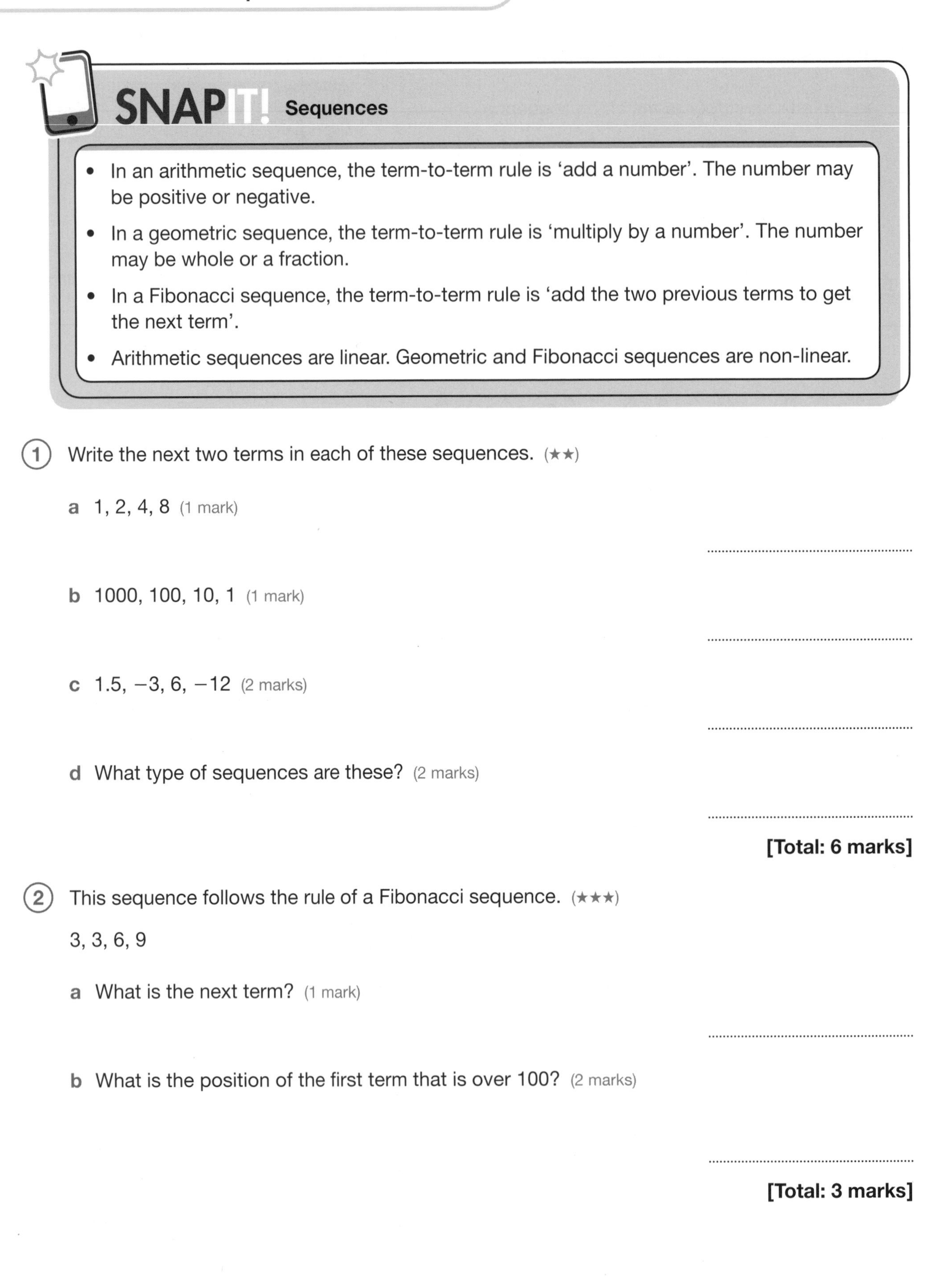

SNAPIT! Sequences

- In an arithmetic sequence, the term-to-term rule is 'add a number'. The number may be positive or negative.
- In a geometric sequence, the term-to-term rule is 'multiply by a number'. The number may be whole or a fraction.
- In a Fibonacci sequence, the term-to-term rule is 'add the two previous terms to get the next term'.
- Arithmetic sequences are linear. Geometric and Fibonacci sequences are non-linear.

1. Write the next two terms in each of these sequences. (★★)

   a 1, 2, 4, 8 (1 mark)

   ....

   b 1000, 100, 10, 1 (1 mark)

   ....

   c 1.5, −3, 6, −12 (2 marks)

   ....

   d What type of sequences are these? (2 marks)

   ....

[Total: 6 marks]

2. This sequence follows the rule of a Fibonacci sequence. (★★★)

   3, 3, 6, 9

   a What is the next term? (1 mark)

   ....

   b What is the position of the first term that is over 100? (2 marks)

   ....

[Total: 3 marks]

(3) Each day a gardener recorded the number of ladybirds on his herbs. (★★★)

Day	Mon	Tue	Wed	Thu	Fri
Number of ladybirds	2	8	32		

a He says, 'If the number of ladybirds continues to grow in the same way, by the end of Friday there will be more than 500 in my herb garden.'

Complete the table to show the number of ladybirds on the next two days.

Is he correct? Show how you get your answer. (2 marks)

....

b If the sequence continues, on what day will there be more than 2000 ladybirds in the herb garden? (1 mark)

....

[Total: 3 marks]

(4) The nth term of a sequence is $\frac{1}{2}n^2$. (★★★★)

a Write down the first three terms of the sequence. (2 marks)

....

b Is 32 in the sequence? Show how you get your answer. (3 marks)

....

[Total: 5 marks]

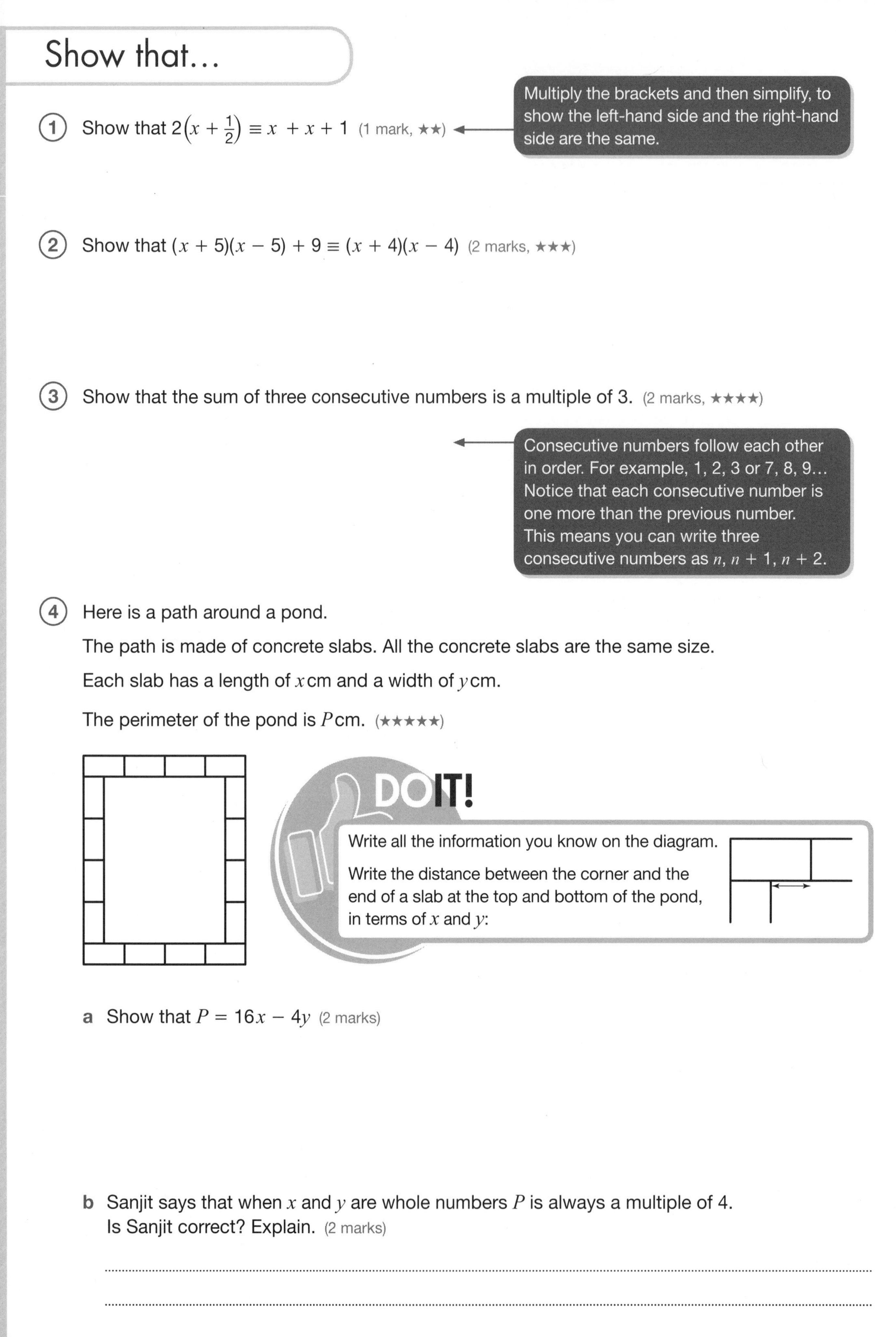

Show that...

① Show that $2\left(x + \frac{1}{2}\right) \equiv x + x + 1$ (1 mark, ★★)

Multiply the brackets and then simplify, to show the left-hand side and the right-hand side are the same.

② Show that $(x + 5)(x - 5) + 9 \equiv (x + 4)(x - 4)$ (2 marks, ★★★)

③ Show that the sum of three consecutive numbers is a multiple of 3. (2 marks, ★★★★)

Consecutive numbers follow each other in order. For example, 1, 2, 3 or 7, 8, 9... Notice that each consecutive number is one more than the previous number. This means you can write three consecutive numbers as n, $n + 1$, $n + 2$.

④ Here is a path around a pond.

The path is made of concrete slabs. All the concrete slabs are the same size.

Each slab has a length of x cm and a width of y cm.

The perimeter of the pond is P cm. (★★★★★)

DO IT!

Write all the information you know on the diagram.

Write the distance between the corner and the end of a slab at the top and bottom of the pond, in terms of x and y:

a Show that $P = 16x - 4y$ (2 marks)

b Sanjit says that when x and y are whole numbers P is always a multiple of 4.
Is Sanjit correct? Explain. (2 marks)

....

....

[Total: 4 marks]

Functions

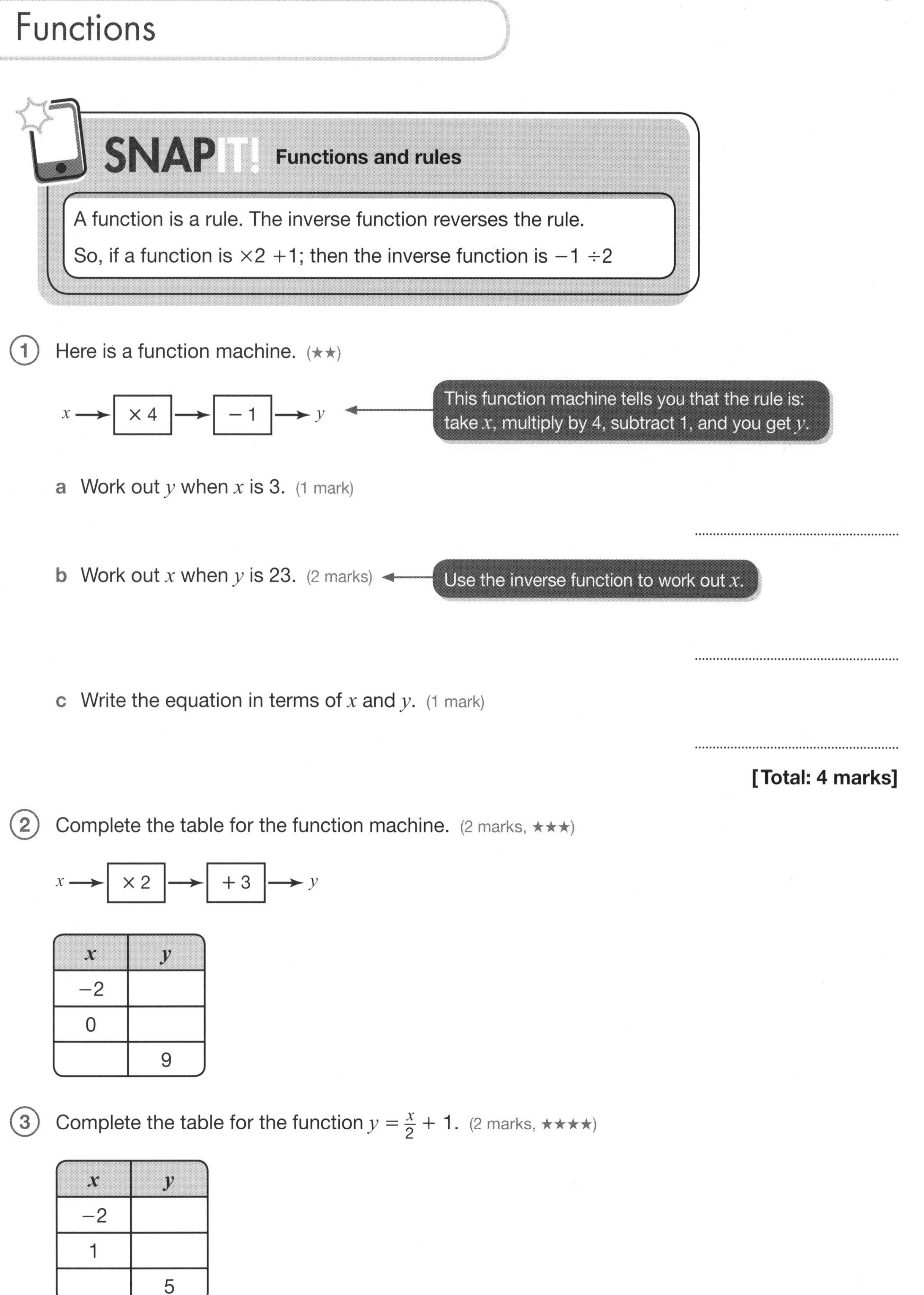

SNAP IT! Functions and rules

A function is a rule. The inverse function reverses the rule.

So, if a function is $\times 2\ +1$; then the inverse function is $-1\ \div 2$

1. Here is a function machine. (★★)

$x \rightarrow \boxed{\times 4} \rightarrow \boxed{-1} \rightarrow y$

This function machine tells you that the rule is: take x, multiply by 4, subtract 1, and you get y.

a Work out y when x is 3. (1 mark)

....

b Work out x when y is 23. (2 marks)

Use the inverse function to work out x.

....

c Write the equation in terms of x and y. (1 mark)

....

[Total: 4 marks]

2. Complete the table for the function machine. (2 marks, ★★★)

$x \rightarrow \boxed{\times 2} \rightarrow \boxed{+3} \rightarrow y$

x	y
−2	
0	
	9

3. Complete the table for the function $y = \frac{x}{2} + 1$. (2 marks, ★★★★)

x	y
−2	
1	
	5

Coordinates and midpoints

1 Here is a graph showing the points *A* and *C*. (★★)

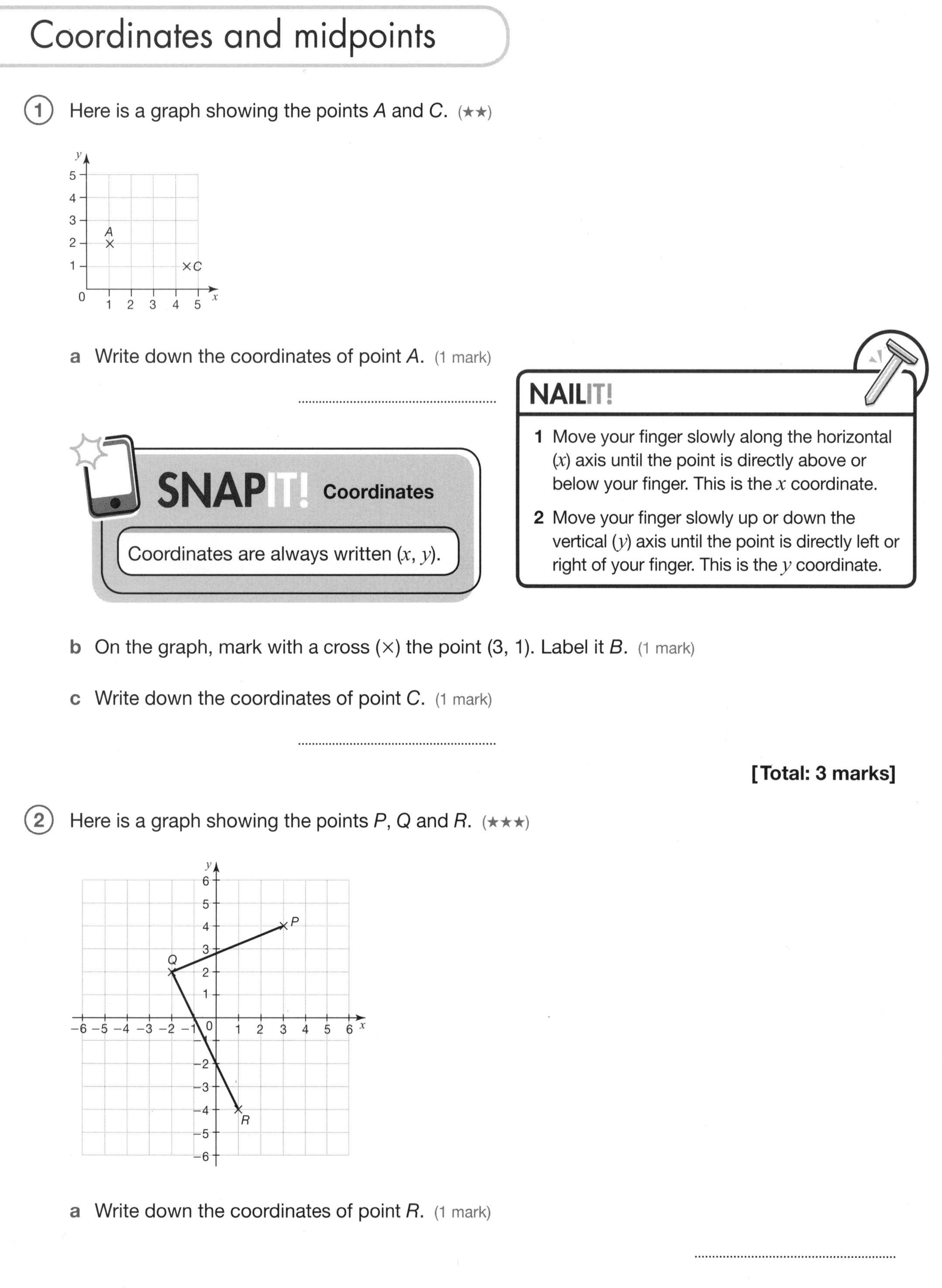

a Write down the coordinates of point *A*. (1 mark)

....

SNAPIT! Coordinates

Coordinates are always written (x, y).

NAILIT!

1. Move your finger slowly along the horizontal (x) axis until the point is directly above or below your finger. This is the x coordinate.
2. Move your finger slowly up or down the vertical (y) axis until the point is directly left or right of your finger. This is the y coordinate.

b On the graph, mark with a cross (×) the point (3, 1). Label it *B*. (1 mark)

c Write down the coordinates of point *C*. (1 mark)

....

[Total: 3 marks]

2 Here is a graph showing the points *P*, *Q* and *R*. (★★★)

a Write down the coordinates of point *R*. (1 mark)

....

b On the graph, mark with a cross (×) a point S, so that the quadrilateral *PQRS* is a parallelogram. Join the points to show the parallelogram. (1 mark)

[Total: 2 marks]

③ Here is a graph showing the triangle X, Y and Z. (★★★★)

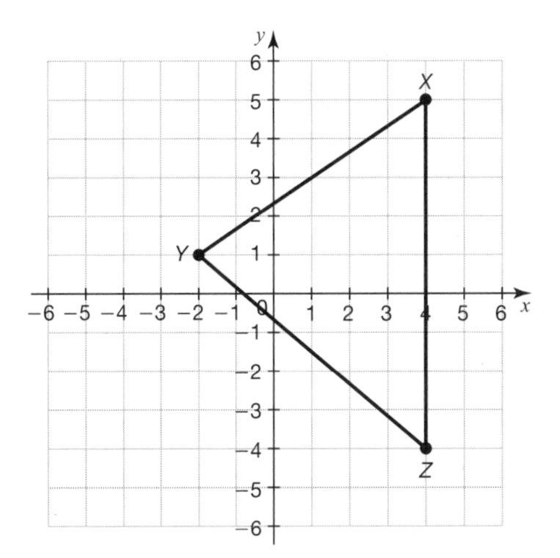

a Write down the midpoint of side XY. (2 marks)

Add the x coordinates for point X and Y. Then halve it to find the midpoint of the x coordinate.

Add the y coordinates for point X and Y. Then halve it to find the midpoint of the y coordinate.

....

b Write down the midpoint of side XZ. (2 marks)

....

c Write down the midpoint of side YZ. (2 marks)

....

[Total: 6 marks]

Straight-line graphs

1 a Complete the table for $y = 2x + 3$ (2 marks, ★)

x	−1	0	1	2
y				

b Draw the graph of $y = 2x + 3$ (2 marks)

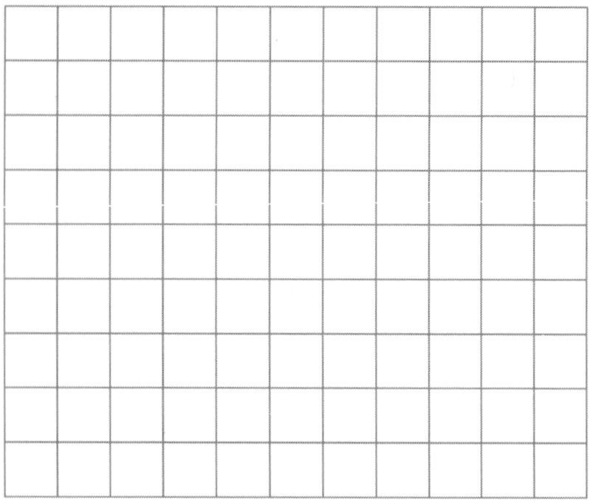

[Total: 4 marks]

2 a Using the axes for graph C, draw the graph of $y = 3x$ for $x = -2$ to 2. Label it A. (2 marks, ★★)

First of all, draw and complete a table with x from −2 to 2.

SNAPIT! Straight-line graph

The equation of a straight-line graph is $y = mx + c$, where m is the gradient (this is how far the graph rises or falls for every 1 unit across).

You can find the gradient by choosing any two points on a graph and working out: $\frac{\text{difference in } y \text{ coordinates}}{\text{difference in } x \text{ coordinates}}$.

c is the y-intercept (this is where the graph crosses the y-axis).

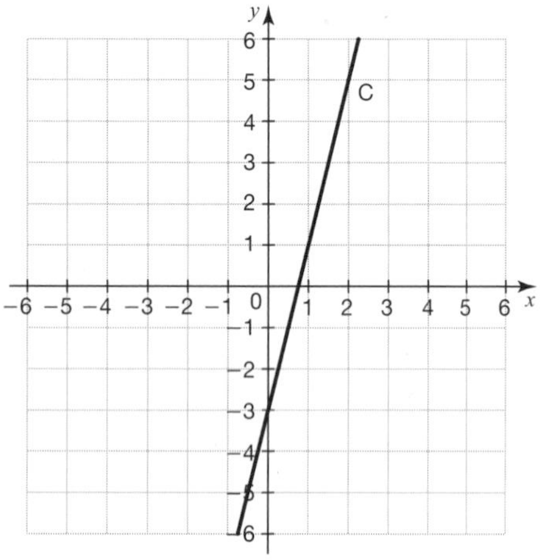

WORKIT!

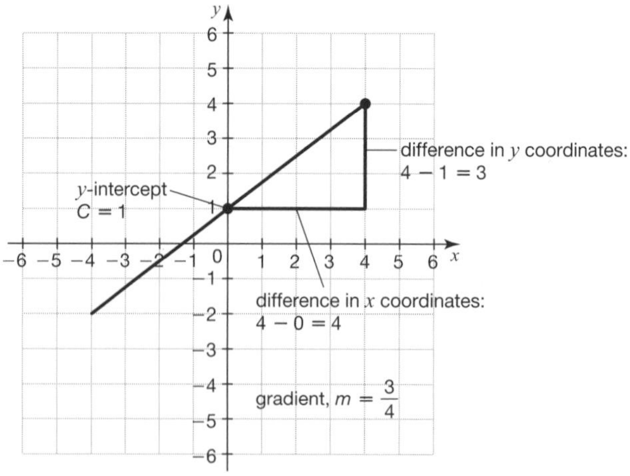

difference in y coordinates: $4 - 1 = 3$

y-intercept $C = 1$

difference in x coordinates: $4 - 0 = 4$

gradient, $m = \frac{3}{4}$

b Using the axes for graph C, draw the graph of $2x + y = -\frac{1}{2}$

Label it B. (2 marks, ★★★★)

First, rearrange the equation to make y the subject.

c Find the equation of the straight-line graph C. (2 marks, ★★★★)

....

[Total: 6 marks]

3. Here are the equations of three straight lines. (★★★★)

A $y = x + 1$ B $y = 2x + 1$ C $y = 2x + 2$

a Which two lines are parallel? Explain. (2 marks)

....

b Which two lines cross the y-axis at the same coordinates? Explain. (2 marks)

....

[Total: 4 marks]

4. A straight line passes through the points with coordinates (3, 2) and (−1, −6).

Find the equation of the line. (3 marks, ★★★★★)

> Find the gradient.
> Substitute the gradient, along with $x = 3$ and $y = 2$, into the equation $y = mx + c$.
> Rearrange the equation to find c.
> Write the equation of the line.

....

SNAPIT! More about straight-line graphs

Straight-line graphs
- are of the form $y = mx + c$
- are parallel if they have the same gradient, m
- cross the y-axis at the same coordinates if they have the same y-intercept, c.

5. Here is the graph of $y = 2x - 4$ (★)

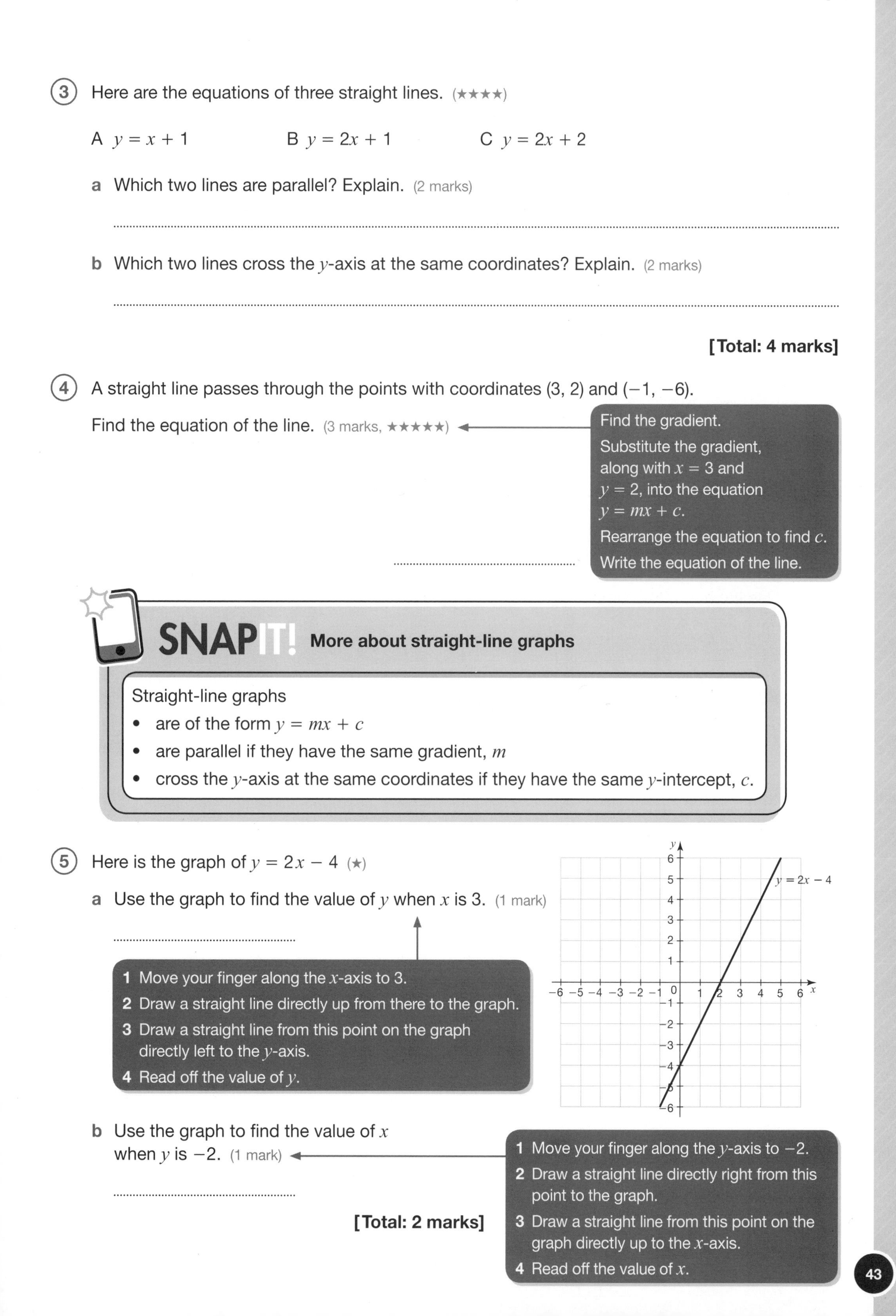

a Use the graph to find the value of y when x is 3. (1 mark)

....

> 1 Move your finger along the x-axis to 3.
> 2 Draw a straight line directly up from there to the graph.
> 3 Draw a straight line from this point on the graph directly left to the y-axis.
> 4 Read off the value of y.

b Use the graph to find the value of x when y is −2. (1 mark)

....

> 1 Move your finger along the y-axis to −2.
> 2 Draw a straight line directly right from this point to the graph.
> 3 Draw a straight line from this point on the graph directly up to the x-axis.
> 4 Read off the value of x.

[Total: 2 marks]

6 Here is the graph of $2x + y = 3$ (★★★)

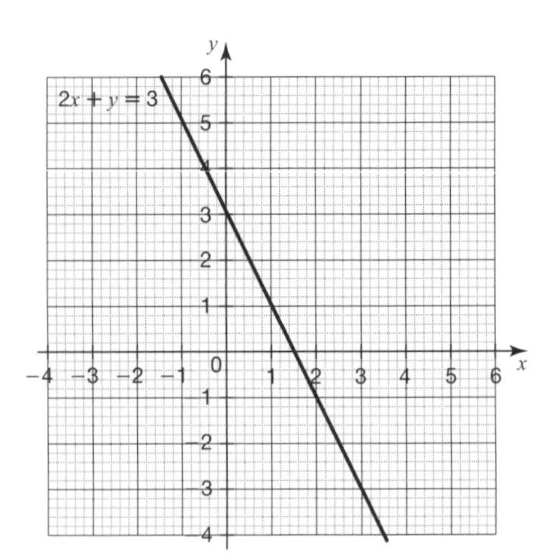

a Use the graph to find the value of x when y is -2. (1 mark)

....

b Use the graph to find the value of y when x is -0.5. (1 mark)

....

c Use the graph to find the value of y when x is 1.2. (1 mark)

....

[Total: 3 marks]

7 Here is the graph of $y = 3x - 1$ (★★★)

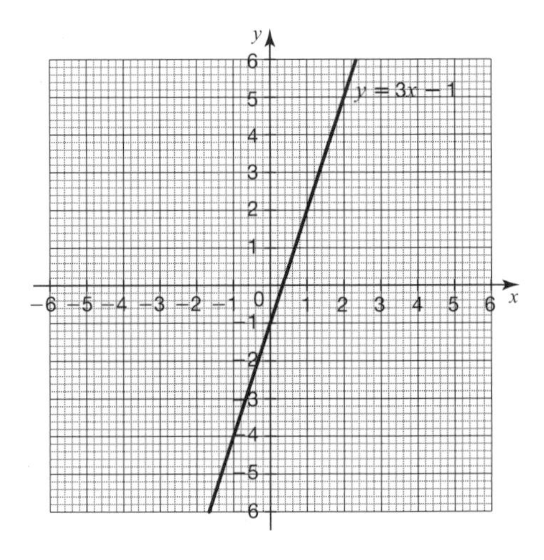

a On the same axis, draw the graph of $y = -3$ (1 mark)

b Use the graph to estimate a solution to the equation $3x - 1 = -3$ (1 mark)

....

What is the value of x where the graphs intersect (cross)?

[Total: 2 marks]

8 Draw the graph of $y = 4x - 3$ for values of x from -5 to 5.

Use the graph to estimate a solution to the equation $4x - 3 = 2$ (3 marks, ★★★★★)

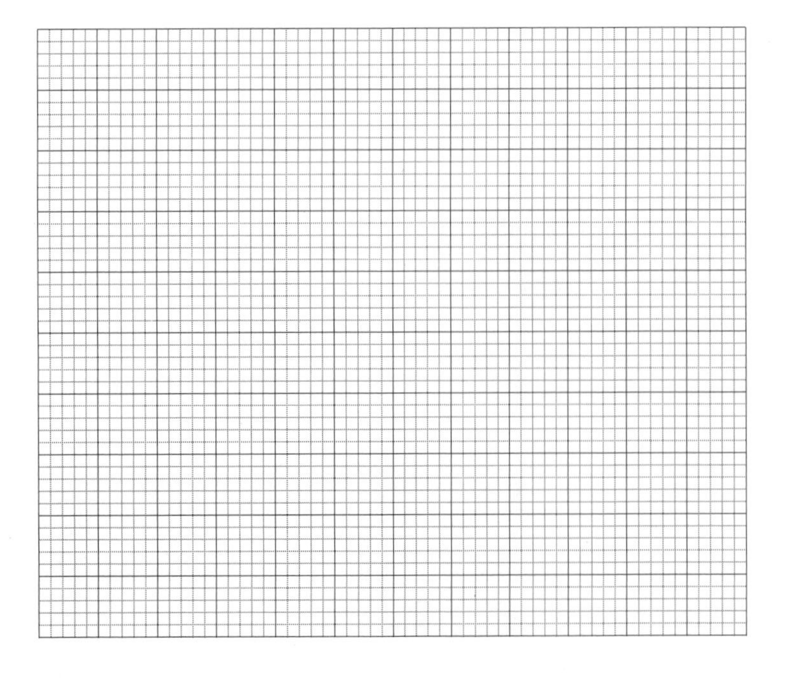

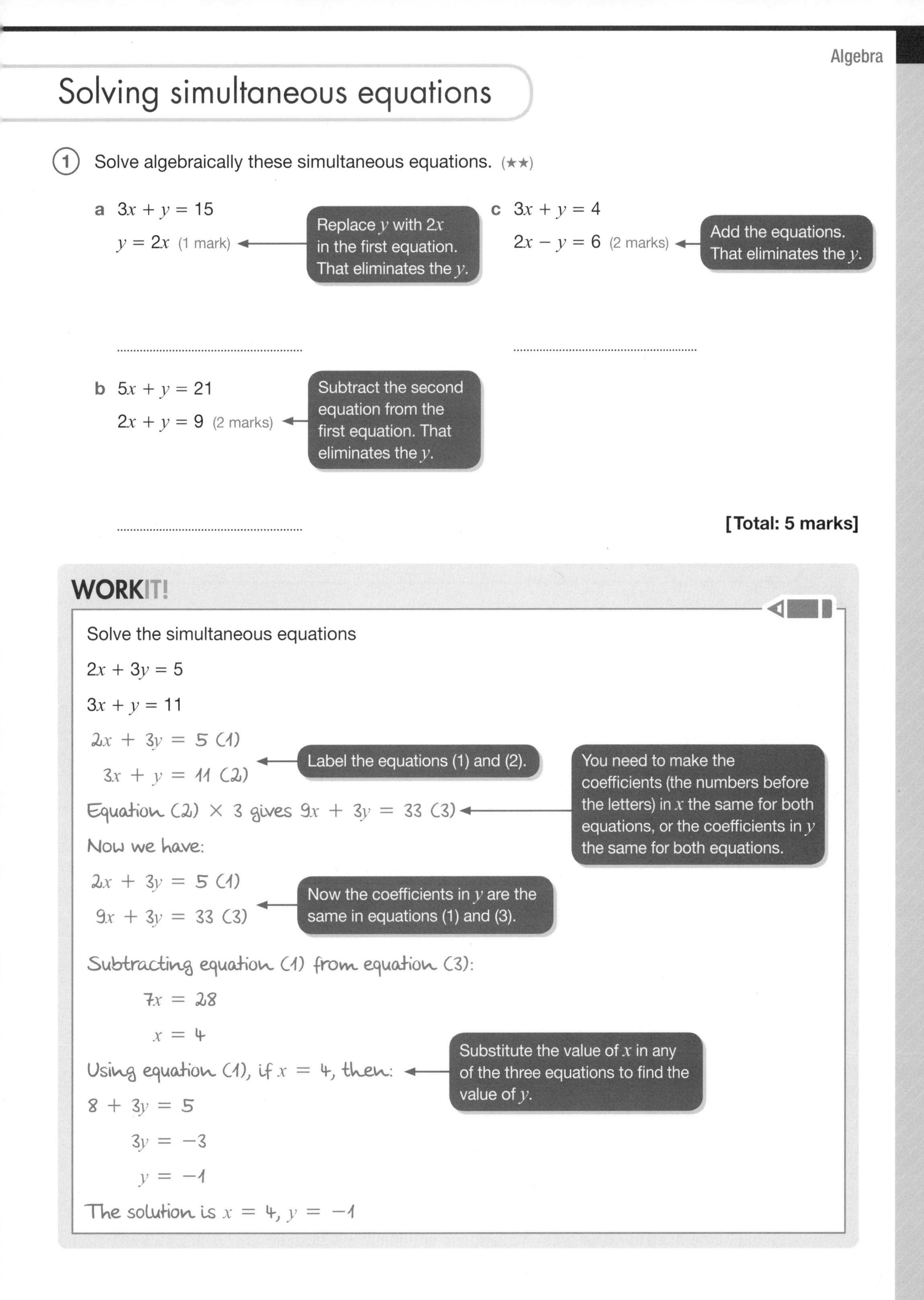

Solving simultaneous equations

1. Solve algebraically these simultaneous equations. (★★)

a $3x + y = 15$

$y = 2x$ (1 mark)

Replace y with $2x$ in the first equation. That eliminates the y.

....

b $5x + y = 21$

$2x + y = 9$ (2 marks)

Subtract the second equation from the first equation. That eliminates the y.

....

c $3x + y = 4$

$2x - y = 6$ (2 marks)

Add the equations. That eliminates the y.

....

[Total: 5 marks]

WORKIT!

Solve the simultaneous equations

$2x + 3y = 5$

$3x + y = 11$

$2x + 3y = 5$ (1)

$3x + y = 11$ (2)

Label the equations (1) and (2).

Equation (2) × 3 gives $9x + 3y = 33$ (3)

You need to make the coefficients (the numbers before the letters) in x the same for both equations, or the coefficients in y the same for both equations.

Now we have:

$2x + 3y = 5$ (1)

$9x + 3y = 33$ (3)

Now the coefficients in y are the same in equations (1) and (3).

Subtracting equation (1) from equation (3):

$7x = 28$

$x = 4$

Using equation (1), if $x = 4$, then:

Substitute the value of x in any of the three equations to find the value of y.

$8 + 3y = 5$

$3y = -3$

$y = -1$

The solution is $x = 4$, $y = -1$

(2) Solve algebraically these simultaneous equations. (★★★★)

a $2x + 2y = 14$

$3x + y = 11$ (3 marks)

....

b $4x - 2y = 2$

$2x - 3y = 7$ (3 marks)

....

c $2x + 3y = 20$

$3x + 2y = 15$ (4 marks)

....

[Total: 10 marks]

NAILIT!

Sometimes it is necessary to multiply both the first and the second equation by a number so that the coefficients in x or the coefficients in y in both equations are the same.

(3) **a** Complete this table for $x + y = 2$

x	0	
y		0

Complete this table for $2x - y = 1$ (2 marks, ★★★★)

x	0	
y		0

b Solve graphically these simultaneous equations. (1 mark, ★★★★★)

$x + y = 2$

$2x - y = 1$

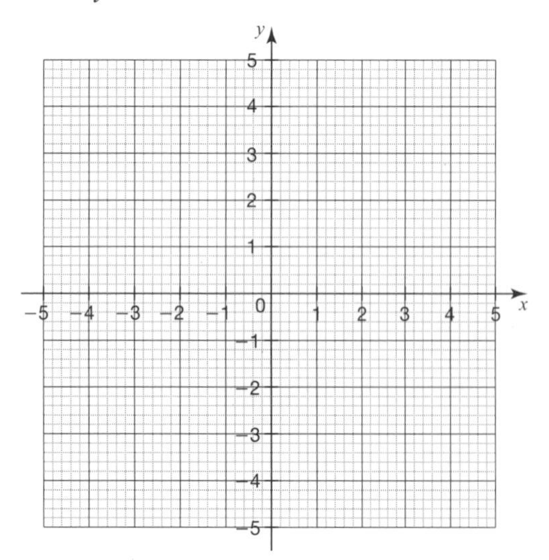

NAILIT!

To solve simultaneous equations graphically, draw the graph of each equation and find the values of x and y where they intersect (cross).

c Use the same axes to solve graphically these simultaneous equations. (1 mark, ★★★★★)

$2x - y = 1$

$-x + 2y = 4$

[Total: 4 marks]

Quadratic graphs

1. Here are five graphs. (★)

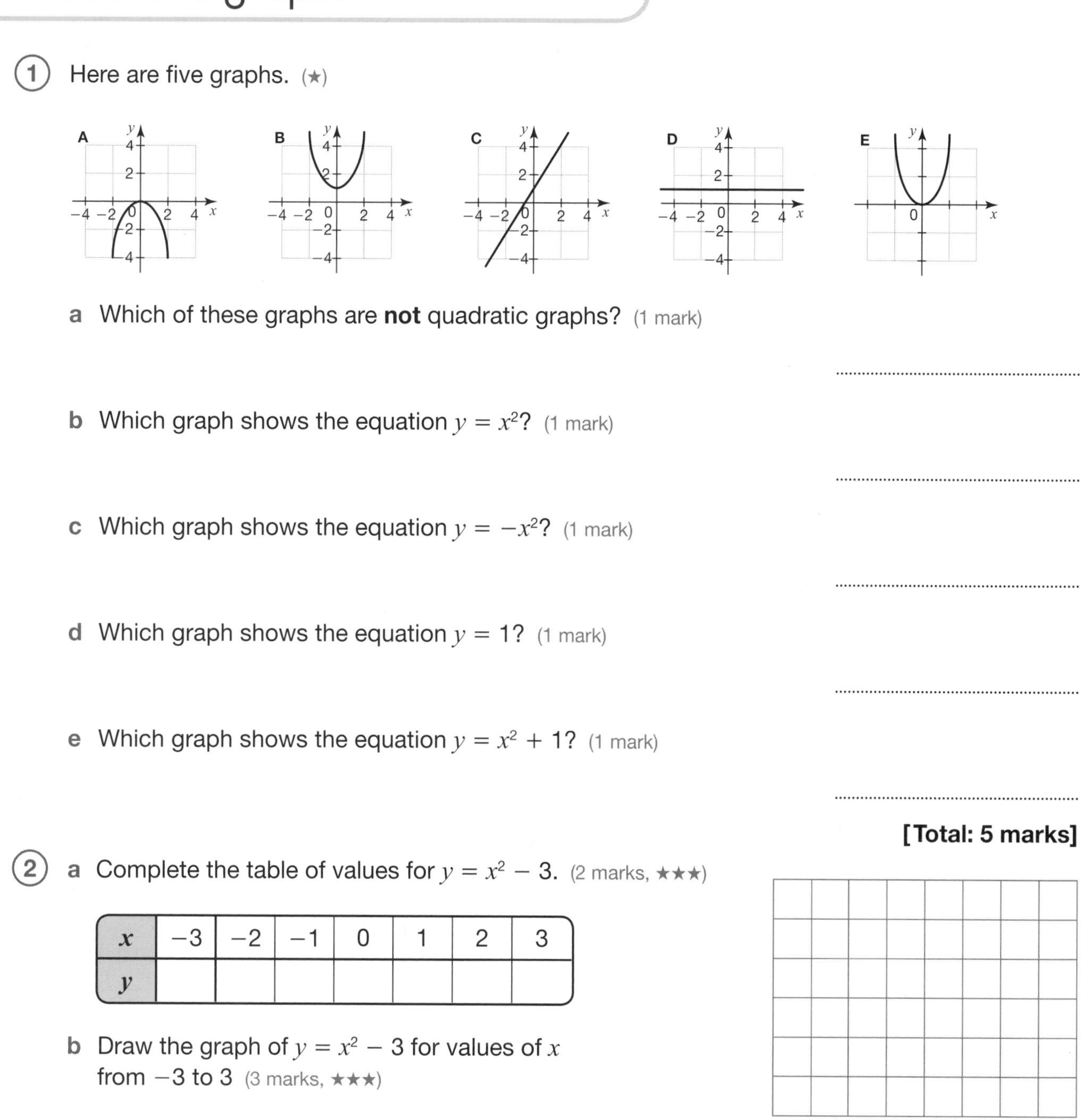

a Which of these graphs are **not** quadratic graphs? (1 mark)

....

b Which graph shows the equation $y = x^2$? (1 mark)

....

c Which graph shows the equation $y = -x^2$? (1 mark)

....

d Which graph shows the equation $y = 1$? (1 mark)

....

e Which graph shows the equation $y = x^2 + 1$? (1 mark)

....

[Total: 5 marks]

2. **a** Complete the table of values for $y = x^2 - 3$. (2 marks, ★★★)

x	−3	−2	−1	0	1	2	3
y							

b Draw the graph of $y = x^2 - 3$ for values of x from −3 to 3 (3 marks, ★★★)

c Write down the equation of the line of symmetry. (1 mark, ★★★)

.... ← The line of symmetry is at $x =$ ☐

d Write down the coordinates of the turning point. (1 mark, ★★★)

....

[Total: 7 marks]

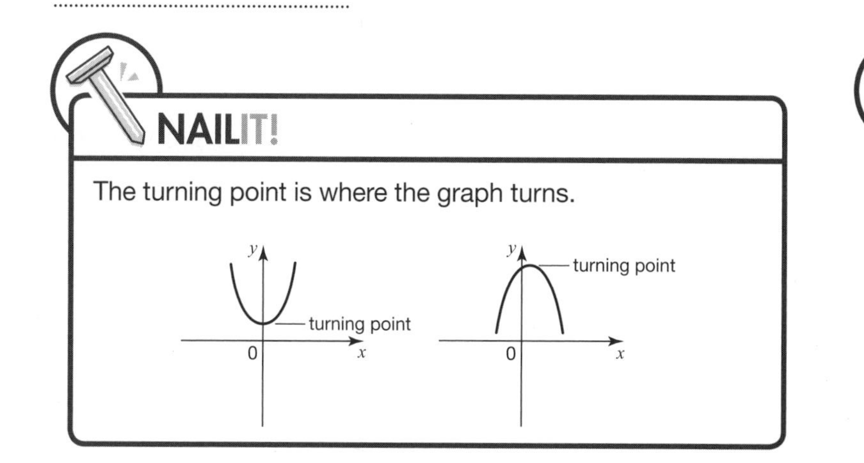

When drawing quadratic graphs
- use a sharp pencil
- plot each point with a cross (×)
- draw a smooth curve through **all** the points.

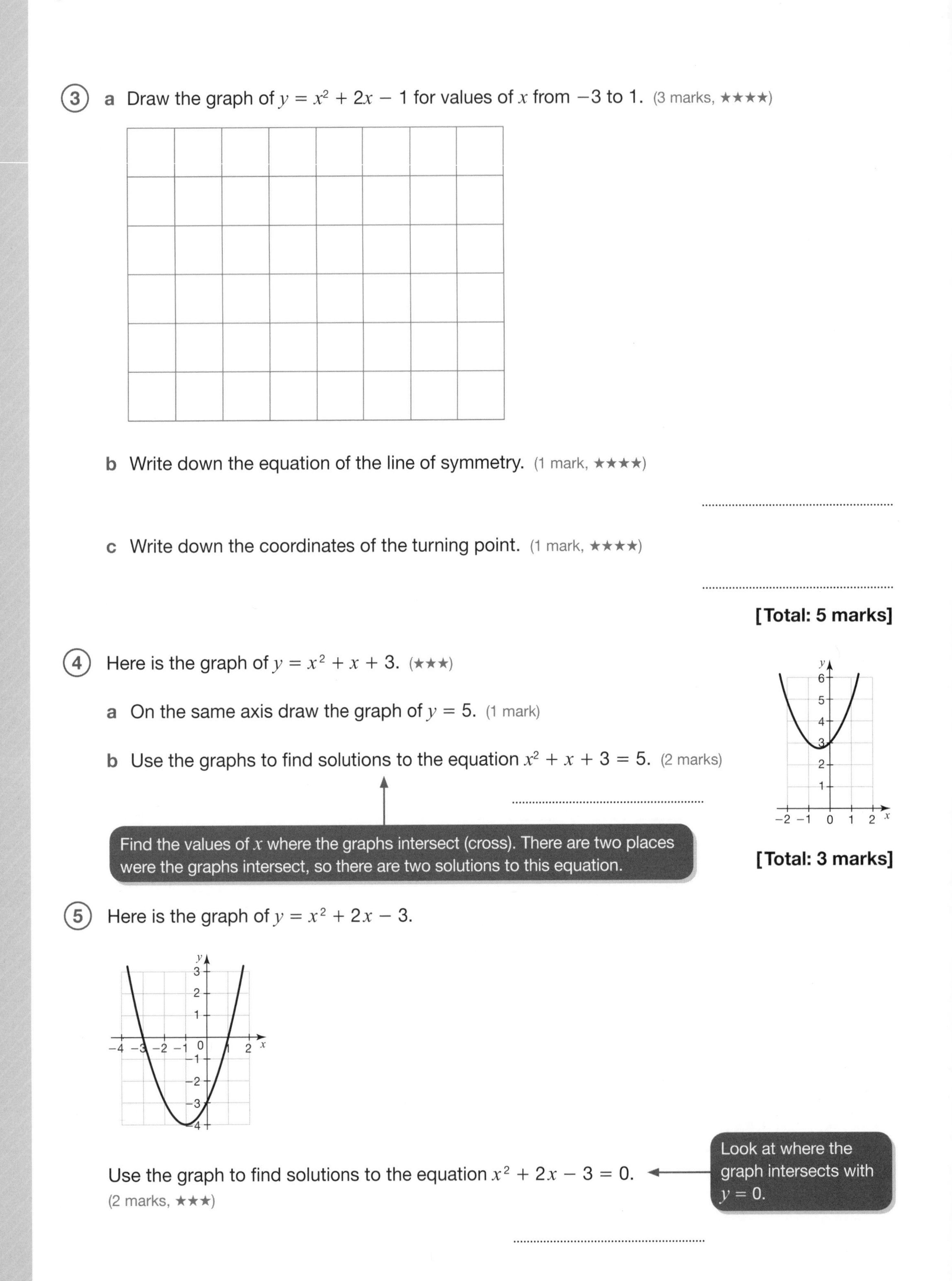

③ **a** Draw the graph of $y = x^2 + 2x - 1$ for values of x from -3 to 1. (3 marks, ★★★★)

b Write down the equation of the line of symmetry. (1 mark, ★★★★)

....

c Write down the coordinates of the turning point. (1 mark, ★★★★)

....

[Total: 5 marks]

④ Here is the graph of $y = x^2 + x + 3$. (★★★)

a On the same axis draw the graph of $y = 5$. (1 mark)

b Use the graphs to find solutions to the equation $x^2 + x + 3 = 5$. (2 marks)

....

Find the values of x where the graphs intersect (cross). There are two places were the graphs intersect, so there are two solutions to this equation.

[Total: 3 marks]

⑤ Here is the graph of $y = x^2 + 2x - 3$.

Use the graph to find solutions to the equation $x^2 + 2x - 3 = 0$. (2 marks, ★★★)

Look at where the graph intersects with $y = 0$.

....

(6) **a** Draw the graph of $y = x^2 + 3x + 1$ for values of x from -3 to 0. (3 marks, ★★★★★)

Draw and complete a table of values of $y = x^2 + 3x + 1$ for values of x from -3 to 0.

x	-3	-2	-1	0
y				

b Use the graph to find solutions to the equation $x^2 + 3x + 1 = -1$. (2 marks, ★★★★★)

....

c Use the graph to estimate solutions to the equation $x^2 + 3x + 1 = 0$. (2 marks, ★★★★★)

....

[Total: 7 marks]

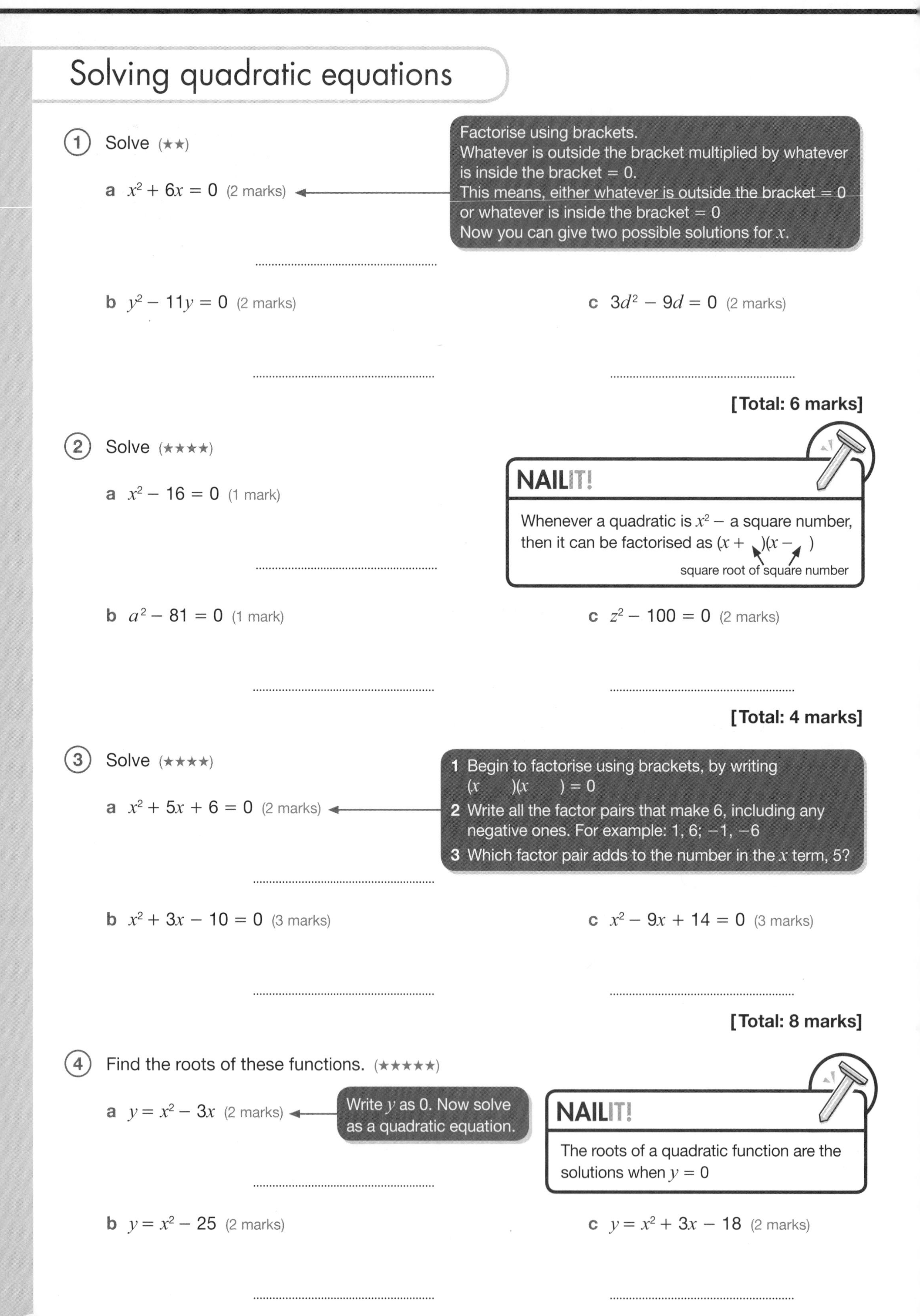

Solving quadratic equations

1. Solve (★★)

Factorise using brackets.
Whatever is outside the bracket multiplied by whatever is inside the bracket = 0.
This means, either whatever is outside the bracket = 0 or whatever is inside the bracket = 0
Now you can give two possible solutions for x.

a $x^2 + 6x = 0$ (2 marks)

....

b $y^2 - 11y = 0$ (2 marks)

....

c $3d^2 - 9d = 0$ (2 marks)

....

[Total: 6 marks]

2. Solve (★★★★)

NAILIT!

Whenever a quadratic is x^2 – a square number, then it can be factorised as $(x + \ )(x - \ )$

square root of square number

a $x^2 - 16 = 0$ (1 mark)

....

b $a^2 - 81 = 0$ (1 mark)

....

c $z^2 - 100 = 0$ (2 marks)

....

[Total: 4 marks]

3. Solve (★★★★)

1. Begin to factorise using brackets, by writing $(x \quad)(x \quad) = 0$
2. Write all the factor pairs that make 6, including any negative ones. For example: 1, 6; −1, −6
3. Which factor pair adds to the number in the x term, 5?

a $x^2 + 5x + 6 = 0$ (2 marks)

....

b $x^2 + 3x - 10 = 0$ (3 marks)

....

c $x^2 - 9x + 14 = 0$ (3 marks)

....

[Total: 8 marks]

4. Find the roots of these functions. (★★★★★)

Write y as 0. Now solve as a quadratic equation.

NAILIT!

The roots of a quadratic function are the solutions when $y = 0$

a $y = x^2 - 3x$ (2 marks)

....

b $y = x^2 - 25$ (2 marks)

....

c $y = x^2 + 3x - 18$ (2 marks)

....

[Total: 6 marks]

Cubic and reciprocal graphs

1. Here are five graphs. (★)

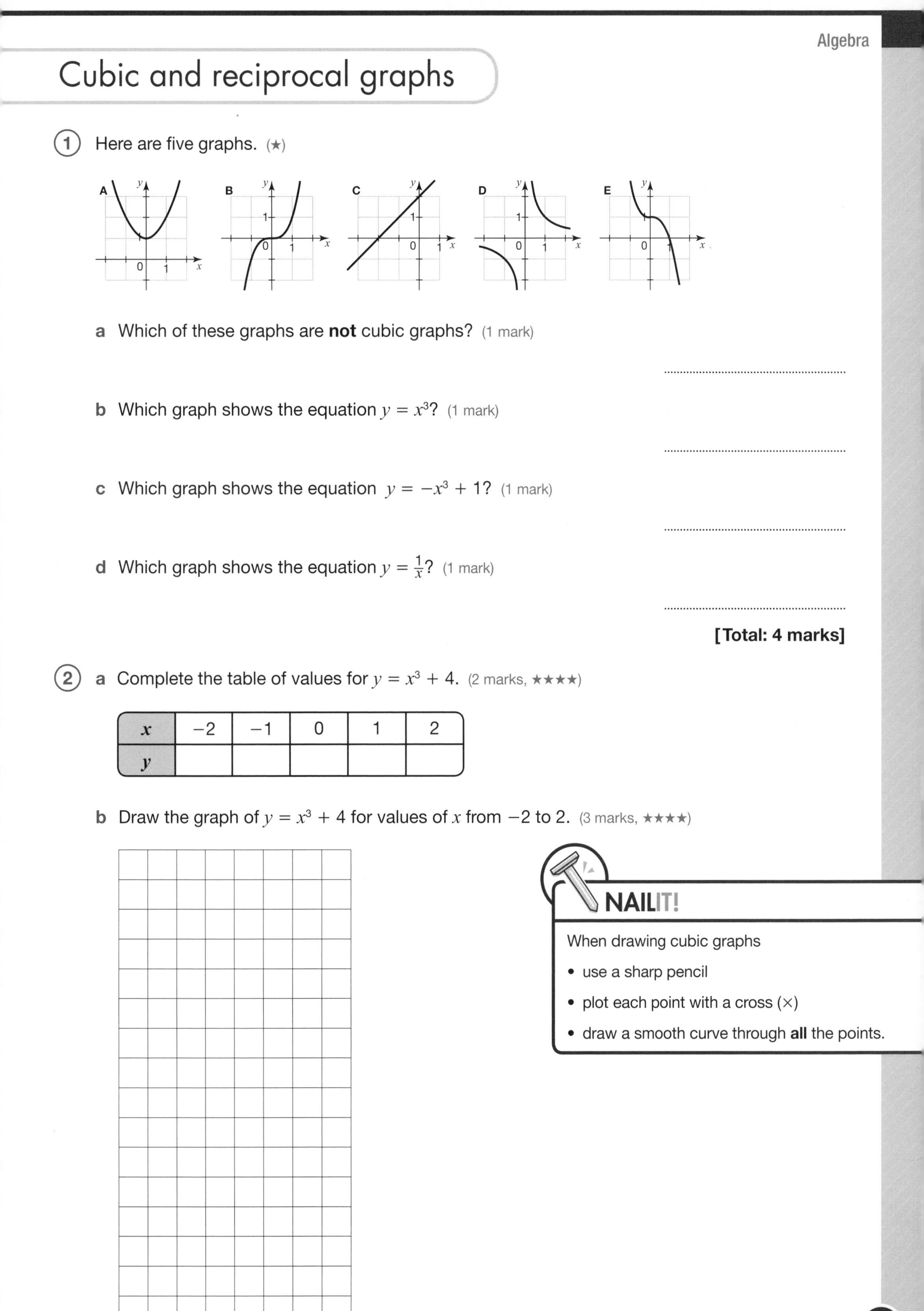

a Which of these graphs are **not** cubic graphs? (1 mark)

....

b Which graph shows the equation $y = x^3$? (1 mark)

....

c Which graph shows the equation $y = -x^3 + 1$? (1 mark)

....

d Which graph shows the equation $y = \frac{1}{x}$? (1 mark)

....

[Total: 4 marks]

2. **a** Complete the table of values for $y = x^3 + 4$. (2 marks, ★★★★)

x	−2	−1	0	1	2
y					

b Draw the graph of $y = x^3 + 4$ for values of x from −2 to 2. (3 marks, ★★★★)

NAILIT!

When drawing cubic graphs

- use a sharp pencil
- plot each point with a cross (×)
- draw a smooth curve through **all** the points.

[Total: 5 marks]

(3) Here is the graph of $y = x^3 - 8$ (★★★★)

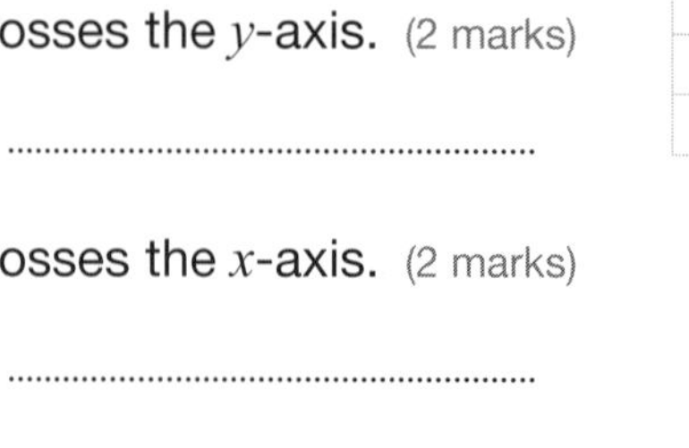

a What type of graph is it? (1 mark)

....

b Write the coordinates where the curve crosses the y-axis. (2 marks)

....

c Write the coordinates where the curve crosses the x-axis. (2 marks)

....

[Total: 5 marks]

(4) **a** Complete the table of values for $y = \frac{1}{x} + 1$. (3 marks, ★★★★★)

x	-3	-2	-1	$-\frac{1}{2}$	$\frac{1}{2}$	1	2	3
y	$\frac{2}{3}$			-1				

b Draw the graph of $y = \frac{1}{x} + 1$ for values of x from -3 to 3. (3 marks, ★★★★★)

[Total: 6 marks]

Drawing and interpreting real-life graphs

(1) The graph shows the cost of a taxi in New York, in US dollars. (★★★★)

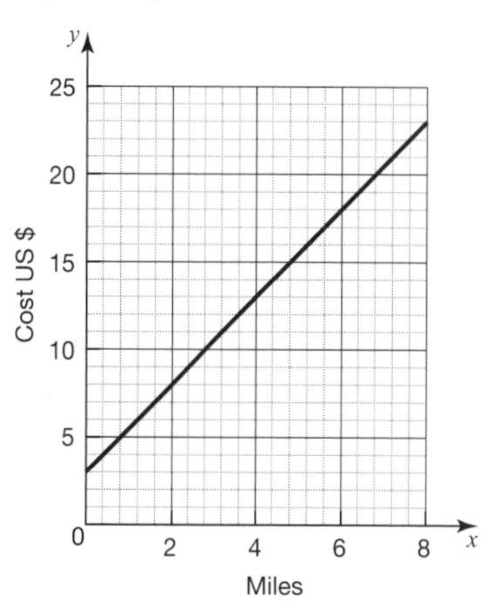

a How much is the initial charge? (1 mark)

....

b How much is the charge per mile? (2 marks)

The gradient of the graph, $\frac{\text{difference in } y \text{ coordinates}}{\text{difference in } x \text{ coordinates}}$ gives the charge per mile.

....

c You can use the information in the table to convert between UK pounds and US dollars.

UK £	0	5	20	50
US $	0	6	24	60

Use this information to draw a conversion graph. (2 marks)

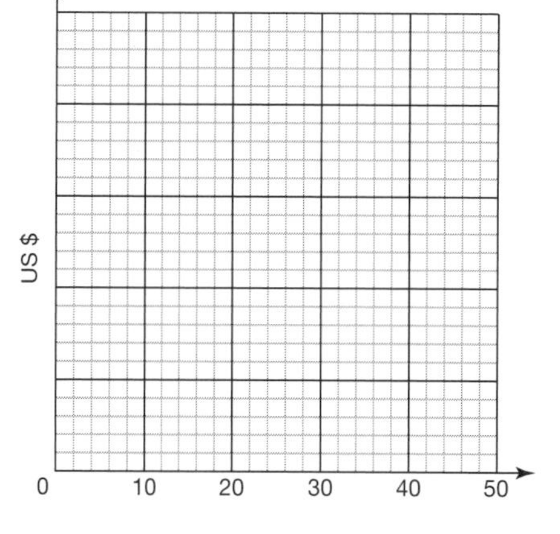

d How far can you travel in a New York taxi for £15? (2 marks)

You will need to use both graphs.

....

[Total: 7 marks]

(2) The graph shows the height of a ball, in metres, against time, in seconds. (★★★)

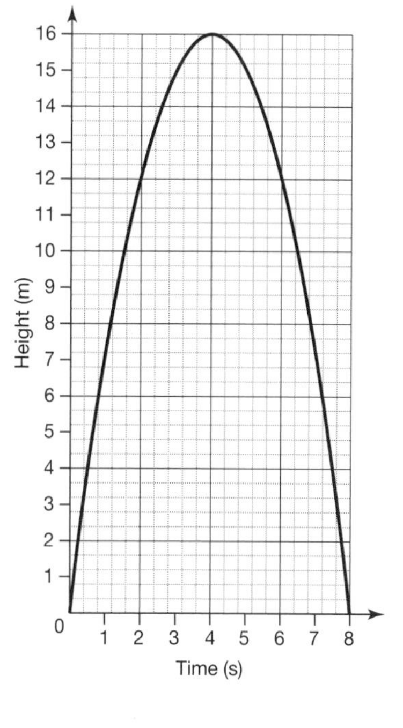

a What is the maximum height the ball reaches? (1 mark)

....

b At how many seconds does the ball reach its maximum height? (1 mark)

....

c After how many seconds does the ball hit the ground? (1 mark)

....

d There were two times when the ball was at a height of 12 m. Write down these times. (2 marks)

....

[Total: 5 marks]

3. The graph shows a short part of a cyclist's cycle ride. (★★★★)

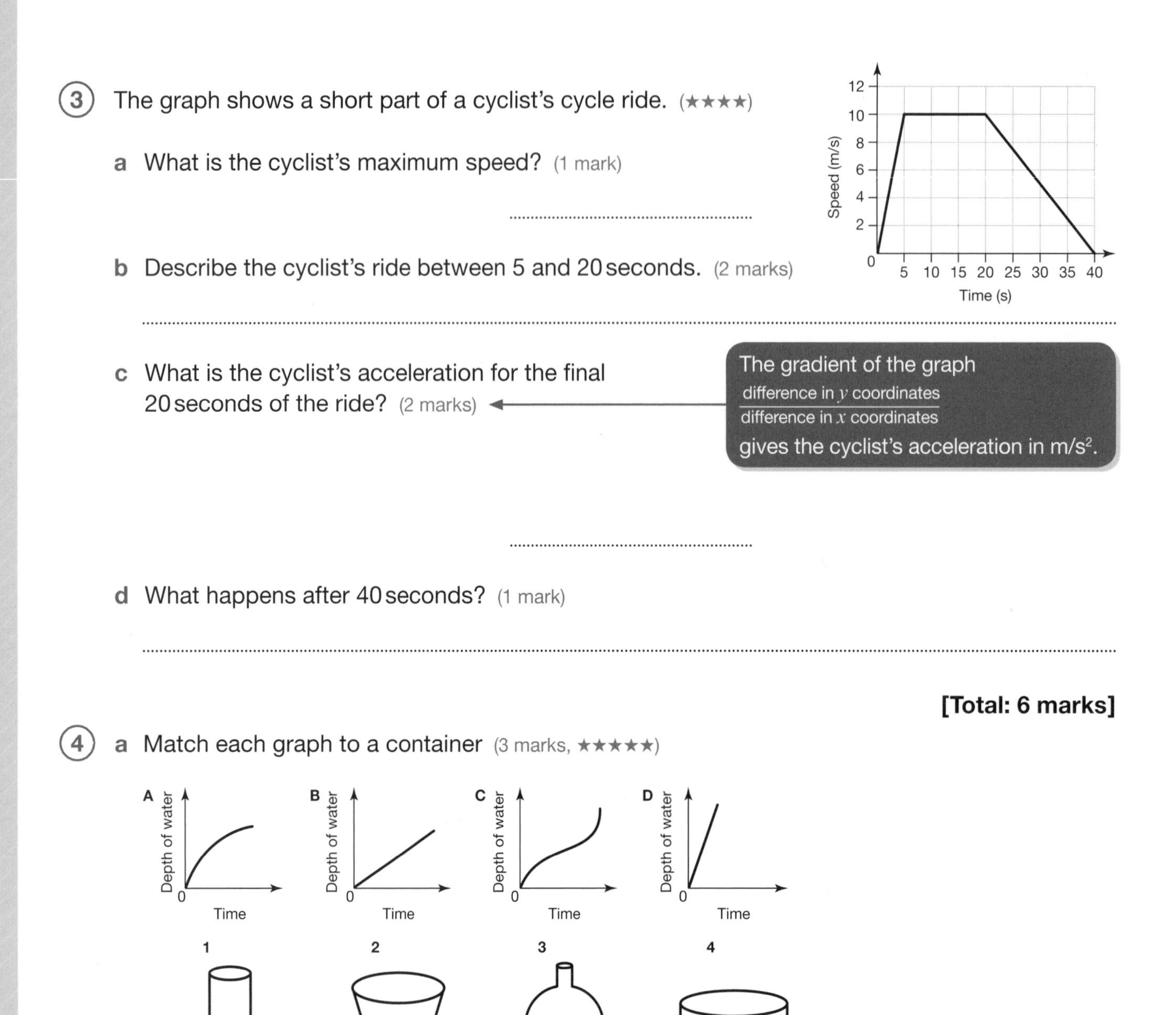

a What is the cyclist's maximum speed? (1 mark)

....

b Describe the cyclist's ride between 5 and 20 seconds. (2 marks)

....

c What is the cyclist's acceleration for the final 20 seconds of the ride? (2 marks)

> The gradient of the graph $\frac{\text{difference in } y \text{ coordinates}}{\text{difference in } x \text{ coordinates}}$ gives the cyclist's acceleration in m/s^2.

....

d What happens after 40 seconds? (1 mark)

....

[Total: 6 marks]

4. **a** Match each graph to a container (3 marks, ★★★★★)

....

....

....

....

b Sketch a depth–time graph for this container. (3 marks, ★★★★★)

[Total: 6 marks]

Ratio, proportion and rates of change
Units of measure

1 Convert these measurements. (★)

a 4 m to cm (1 mark)

....

b 5000 g to kg (1 mark)

....

Ask yourself:
1 What is the conversion factor? (How many cm in 1 m?)
2 Do you expect the number in your answer (in cm) to be more or less than the original number (4)?
3 If more, then multiply by the conversion factor; if less, then divide by the conversion factor.

c 1.5 litres to ml (1 mark)

....

d 8250 m to km (1 mark)

....

[Total: 4 marks]

2 Sally made 6 litres of lemonade. Her friends drank 3500 ml.
How many ml did Sally have left? (2 marks, ★★★)

....

NAIL IT!

Watch out for questions that have two different units of measure (for example litres and millilitres).

Convert one of the measures so you are working in only one type of unit.

3 Luke and his brother Adam left home at the same time.

Luke ran to school in 240 seconds.

Adam ran to school in 3 minutes 47 seconds. (★★★★)

a Which brother arrived at school first?

You must show working to justify your answer. (2 marks)

....

b How long did the first boy at school wait for his brother to arrive? (1 mark)

....

[Total: 3 marks]

4 1 m ≈ 3.2 feet

Ben is 1.25 m; Tom is 4.8 feet. Who is taller?

You must show working to justify your answer. (2 marks, ★★★★★)

....

Ratio

① A child owns 20 books. 12 contain stories and the rest are for colouring. Write the ratio of story to colouring books in its simplest form. (2 marks, ★)

> Write the ratio ☐ : ☐
> Simplify by dividing both numbers by their highest common factor.

....

② Holly has £15. She wants to save and spend it in the ratio 1 : 2.

How much will she spend? (2 marks, ★★)

....

③ Chen has £60.

He wants to save, spend and give some to charity in the ratio 1 : 2 : 3.

How much will he give to charity? (2 marks, ★★★)

....

④ Phil is making green paint.

He plans to mix blue and yellow paint in the ratio 3 : 7 to make 5 litres of green paint.

He has 2 litres of blue paint and 3 litres of yellow paint.

Does he have enough blue and yellow paint to make the green paint?

Show working to justify your answer. (4 marks, ★★★★★)

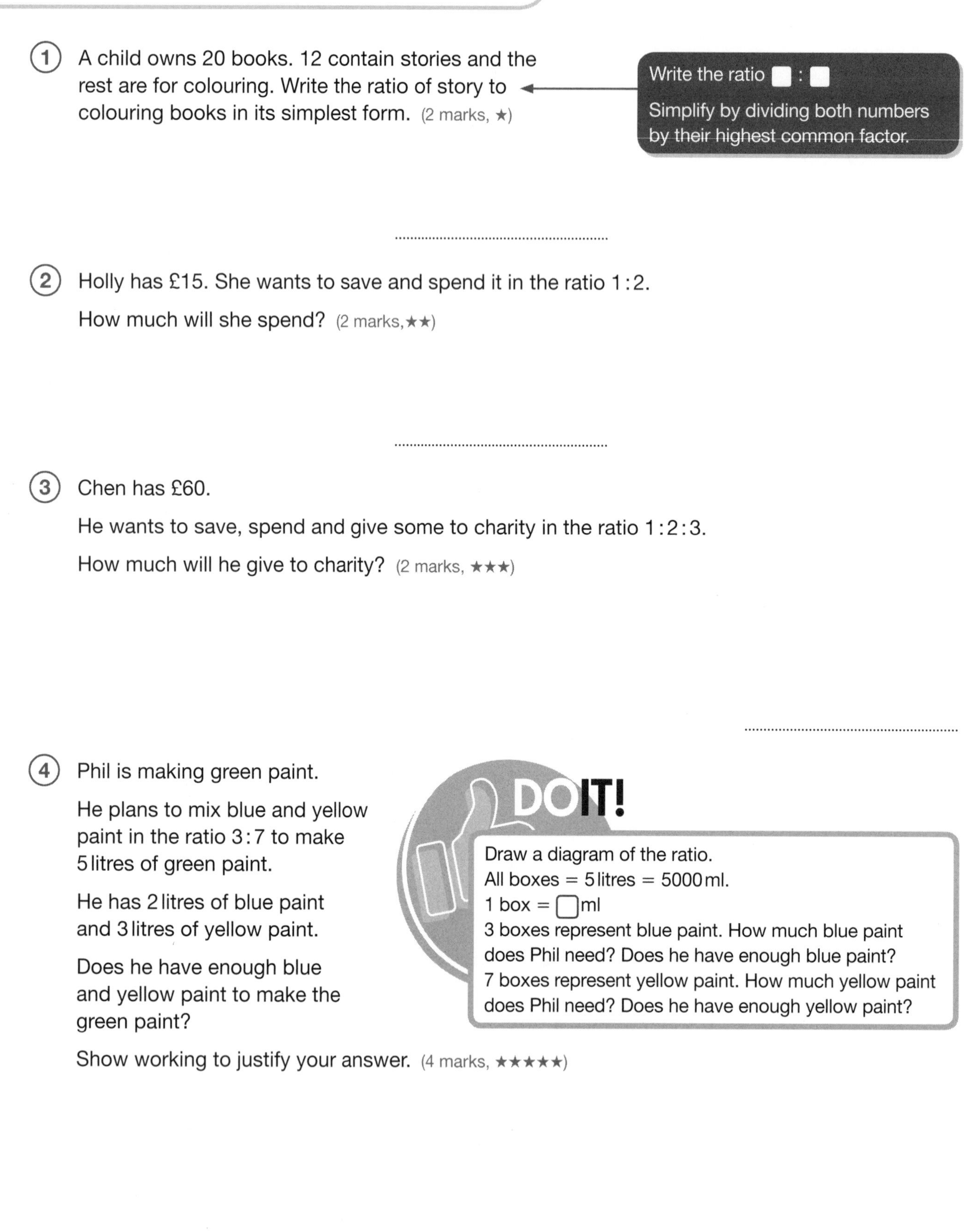

DO IT!

Draw a diagram of the ratio.
All boxes = 5 litres = 5000 ml.
1 box = ☐ ml
3 boxes represent blue paint. How much blue paint does Phil need? Does he have enough blue paint?
7 boxes represent yellow paint. How much yellow paint does Phil need? Does he have enough yellow paint?

....

Scale diagrams and maps

1. Complete these sentences. (1 mark, ★)

   A scale of 1 : 10 000 on a map means

   1 cm on the map is .... cm in real life.

   This means 1 cm on the map is .... m in real life.

2. On a map of a village 1 cm represents 50 m.

   In real life, the village shop is 150 m from the bus stop.

   How far is this, in centimetres, on the map? (1 mark, ★★)

   ....

3. The diagram below represents two trees in a field.

   Tree A ×　　　　　　　Tree B ×

   Scale: 1 cm represents 4 m.

   Work out the distance, in metres, between tree A and tree B. (2 marks, ★★★★)

   ....

4. A scale model of Big Ben in London is 1 : 400

   Big Ben is 96 m high.

   How tall is the model? (2 marks, ★★★★★)

   ....

DO IT!

Write

1 cm on the model is ☐ cm in real life.

1 cm on the model is ☐ m in real life.

☐ cm on the model is 96 m in real life.

Fractions, percentages and proportion

1. Ali and Bess share a pie in the ratio 1 : 3.

   What fraction does Bess receive? (2 marks, ★★)

   ....

SNAPIT! Diagram for ratios

Draw a diagram for ratios. It will help you convert them to fractions.

For example: A : B = 1 : 3

A	B	B	B

2. In a shopping basket, 1 item is a fish, 3 items are fruit, 6 items are tins. (★★)

   a Write the items in the basket as a ratio of fish : fruit : tins. (1 mark)

   ....

   b What fraction of items in the basket is fruit? (1 mark)

   ....

   c What percentage of items in the basket are tins? (1 mark)

   ....

NAILIT!

To write one measure as a fraction of another, both measures must be in the same units.

SNAPIT! Percentage

To write one quantity as a percentage of another:

1. Write as a fraction.
2. Multiply by 100 to convert to a percentage.

[Total: 3 marks]

3. What is 50 g as a fraction of 4 kg? (1 mark, ★★)

   ....

4. A dance school has street dancers, ballet dancers and tap dancers in the ratio 3 : 8 : 14

   What percentage of the dance school do ballet dance? (2 marks, ★★★★)

   ....

Direct proportion

(1) Five concert tickets cost £80. (★)

a How much will one concert ticket cost? (1 mark)

....

b How much will nine concert tickets cost? (1 mark)

....

[Total: 2 marks]

(2) This graph shows the cost of buying packs of 10 colouring pencils. (★)

a What is the cost of 6 packs of 10 pencils?

Give your answer in £. (1 mark)

....

b What is the cost of 1 pencil? (1 mark)

....

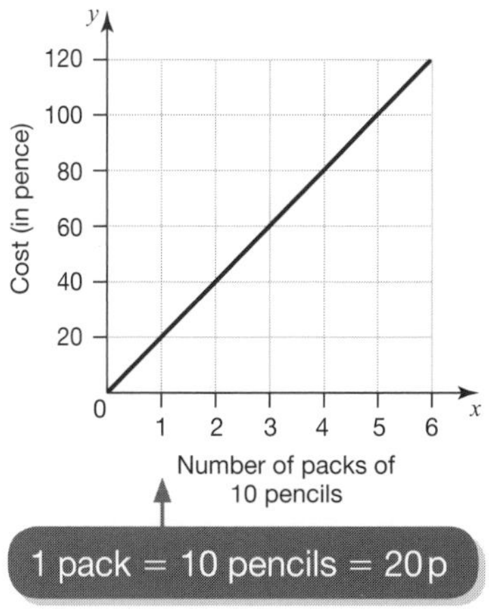

Cost (in pence)

Number of packs of 10 pencils

1 pack = 10 pencils = 20 p

c State **two** ways this graph shows that packs of pencils are in direct proportion to the cost. (2 marks)

....

....

[Total: 4 marks]

(3) A curry recipe is for 4 people. Here are the spices it asks for. (★★★)

1 teaspoon of tumeric

2 teaspoons of chilli powder

$2\frac{1}{2}$ teaspoons of cumin

a Sally is using the recipe to make curry for her class. There are 28 people in Sally's class. How many teaspoons of each spice will she need? (4 marks)

....

....

b 1 teaspoon of chilli powder ≈ 3 g

Sally has 75 g of chilli powder. Does she have enough to make the curry for her class? Show working to justify your answer. (3 marks)

....

....

[Total: 7 marks]

Inverse proportion

① This graph shows competition prize money divided by different numbers of winners. (★)

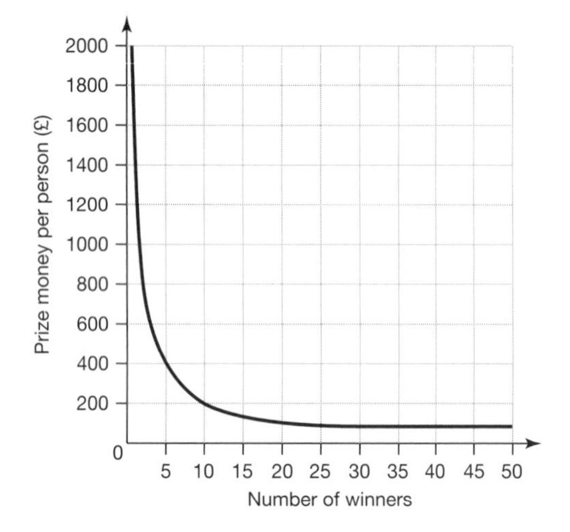

a If there are 5 winners, how much money will they each get? (1 mark)

....

b A winner receives £200. How many other winners are there? (1 mark)

....

c How much is the prize money in total? (1 mark)

....

d Describe the relationship between prize money and number of winners. (1 mark)

....

[Total: 4 marks]

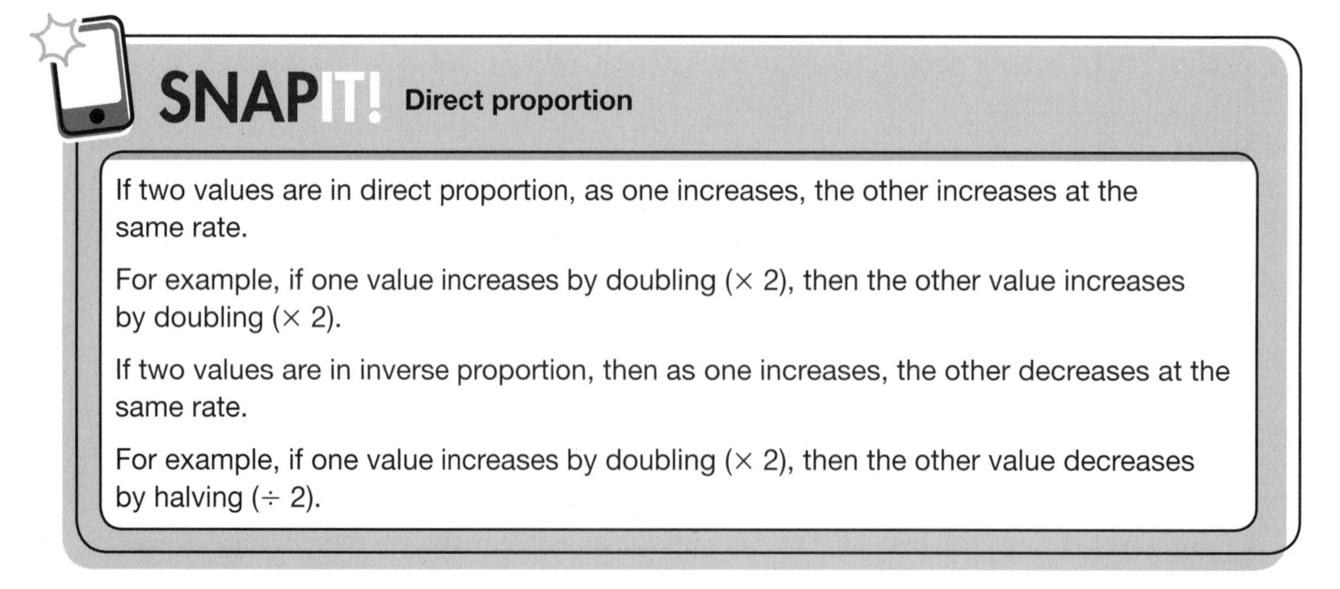

SNAPIT! Direct proportion

If two values are in direct proportion, as one increases, the other increases at the same rate.

For example, if one value increases by doubling (× 2), then the other value increases by doubling (× 2).

If two values are in inverse proportion, then as one increases, the other decreases at the same rate.

For example, if one value increases by doubling (× 2), then the other value decreases by halving (÷ 2).

2. It takes 3 people 2 hours to decorate a room for a party. (★★)

   a How long would it take 1 person? (1 mark)

   Would it take 1 person a longer or shorter amount of time?

   ....

   b How long would it take 12 people? (2 marks)

   It is difficult to work with 3 and 2, as they don't divide nicely into each other. Try changing 2 hours into 120 minutes.

   ....

[Total: 3 marks]

3. It takes a printer 4 minutes to print a 240 page document.

   How long will it take the printer to print a 600 page document? (2 marks, ★★★)

   ....

Working with percentages

1. In a sale, the price of a coat was reduced from £125 to £75.

   Work out the percentage decrease. (1 mark, ★★★)

   ....

SNAP IT! Percentage increase or decrease

To work out percentage increase or decrease:
1. Work out the amount of increase or decrease.
2. Divide by the original value.
3. Multiply by 100.

2. The table shows sales for two different companies in 2006 and 2016. (★★★)

	Company X	Company Y
Sales (£ millions) in 2006	15	
Sales (£ millions) in 2016	24	35

a Work out the percentage increase in sales for Company X between 2006 and 2016. (1 mark)

Decide which numbers in the table you need to find your answer.

....

b Company Y achieved a percentage increase in their sales of 25% between 2006 and 2016.

Work out their sales (£ millions) in 2006. (1 mark)

You know that 125% of the original value in 2006 is £35 million. Use this to work out 1%, then 100%.

....

[Total: 2 marks]

3. Sarah invests £265 in a savings account. It pays 2% interest. (★★★)

a How much will Sarah have in her account after 3 years if the account pays simple interest. (2 marks)

....

b How much will Sarah have in her account after 3 years if the account pays compound interest? (3 marks)

For simple interest: work out how much interest is received after 1 year. Multiply this by 3 and add to £265. For compound interest: work out how much Sarah has after 1 year. Use this amount to find out how much she has after 2 years, and so on.

....

[Total: 5 marks]

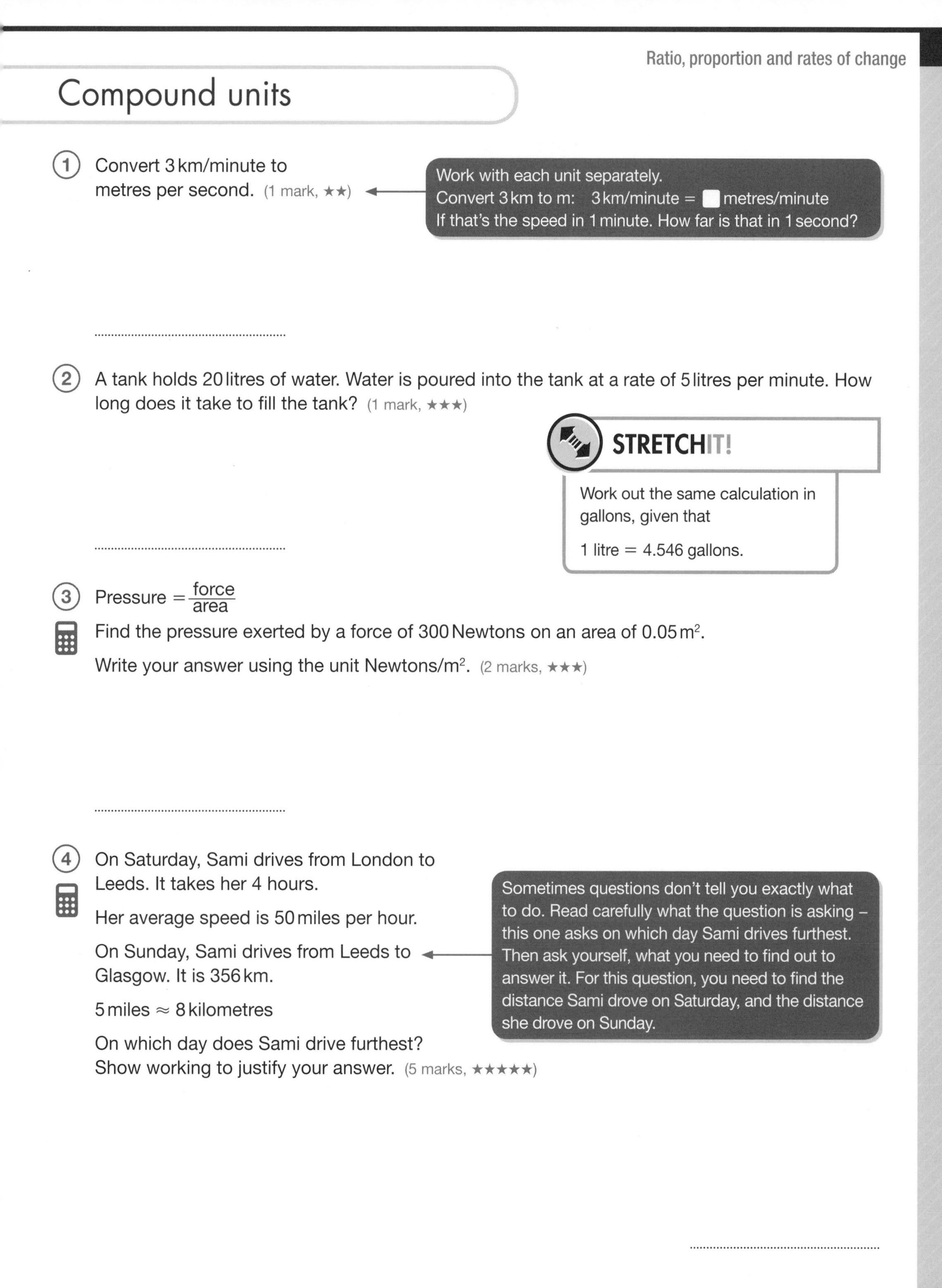

Compound units

1. Convert 3 km/minute to metres per second. (1 mark, ★★)

> Work with each unit separately.
> Convert 3 km to m: 3 km/minute = ☐ metres/minute
> If that's the speed in 1 minute. How far is that in 1 second?

....

2. A tank holds 20 litres of water. Water is poured into the tank at a rate of 5 litres per minute. How long does it take to fill the tank? (1 mark, ★★★)

....

> **STRETCH IT!**
> Work out the same calculation in gallons, given that
> 1 litre = 4.546 gallons.

3. Pressure $= \frac{\text{force}}{\text{area}}$

   Find the pressure exerted by a force of 300 Newtons on an area of 0.05 m^2.

   Write your answer using the unit Newtons/m^2. (2 marks, ★★★)

....

4. On Saturday, Sami drives from London to Leeds. It takes her 4 hours.

   Her average speed is 50 miles per hour.

   On Sunday, Sami drives from Leeds to Glasgow. It is 356 km.

   5 miles $\approx$ 8 kilometres

   On which day does Sami drive furthest?
   Show working to justify your answer. (5 marks, ★★★★★)

> Sometimes questions don't tell you exactly what to do. Read carefully what the question is asking – this one asks on which day Sami drives furthest. Then ask yourself, what you need to find out to answer it. For this question, you need to find the distance Sami drove on Saturday, and the distance she drove on Sunday.

....

Geometry and measures
Measuring and drawing angles

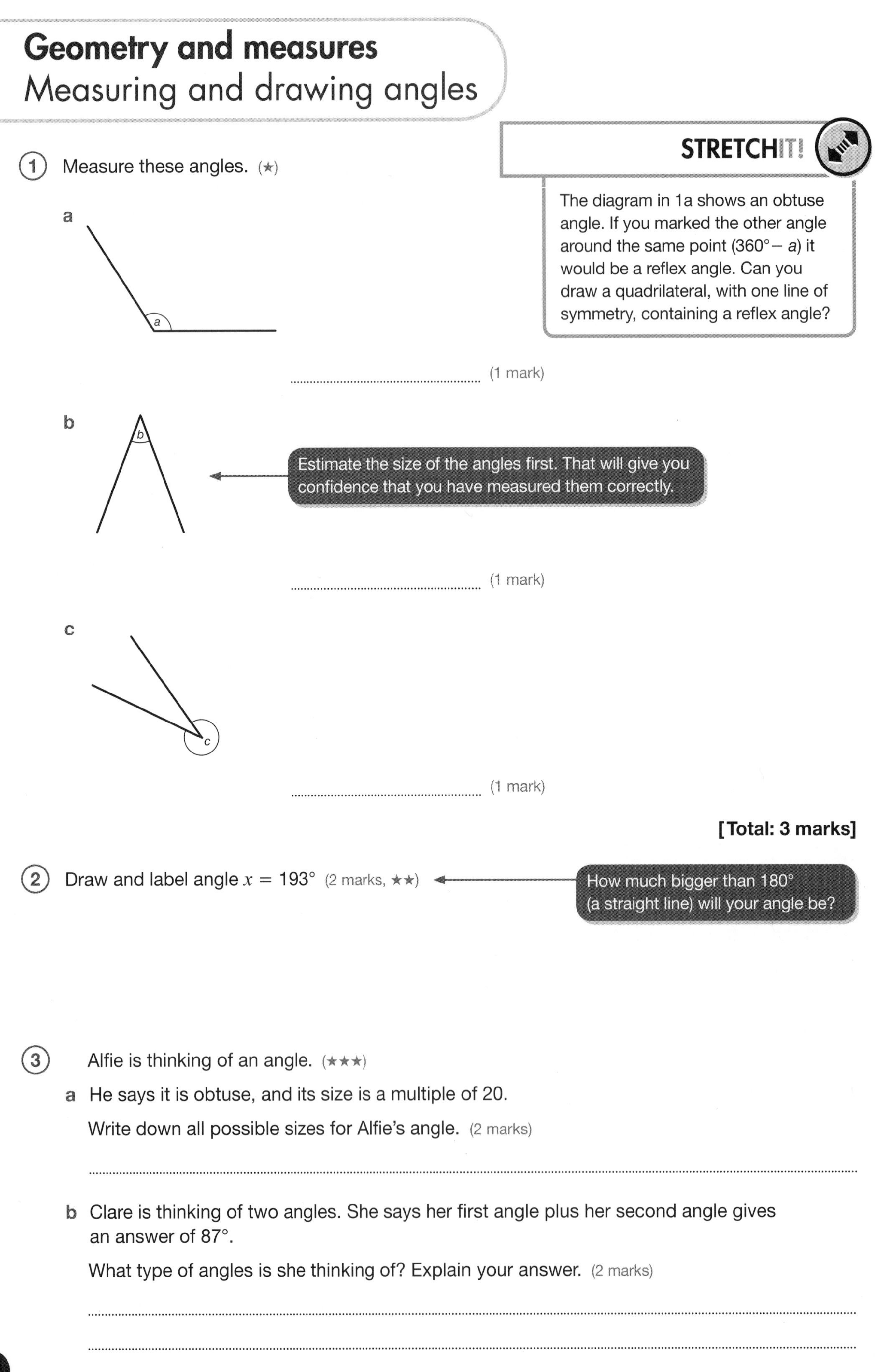

STRETCHIT!

The diagram in 1a shows an obtuse angle. If you marked the other angle around the same point (360° − *a*) it would be a reflex angle. Can you draw a quadrilateral, with one line of symmetry, containing a reflex angle?

1. Measure these angles. (★)

   a

   .... (1 mark)

   b

   Estimate the size of the angles first. That will give you confidence that you have measured them correctly.

   .... (1 mark)

   c

   .... (1 mark)

[Total: 3 marks]

2. Draw and label angle $x = 193°$ (2 marks, ★★)

   How much bigger than 180° (a straight line) will your angle be?

3. Alfie is thinking of an angle. (★★★)

   a He says it is obtuse, and its size is a multiple of 20.

   Write down all possible sizes for Alfie's angle. (2 marks)

   ....

   b Clare is thinking of two angles. She says her first angle plus her second angle gives an answer of 87°.

   What type of angles is she thinking of? Explain your answer. (2 marks)

   ....

   ....

[Total: 4 marks]

Using the properties of angles

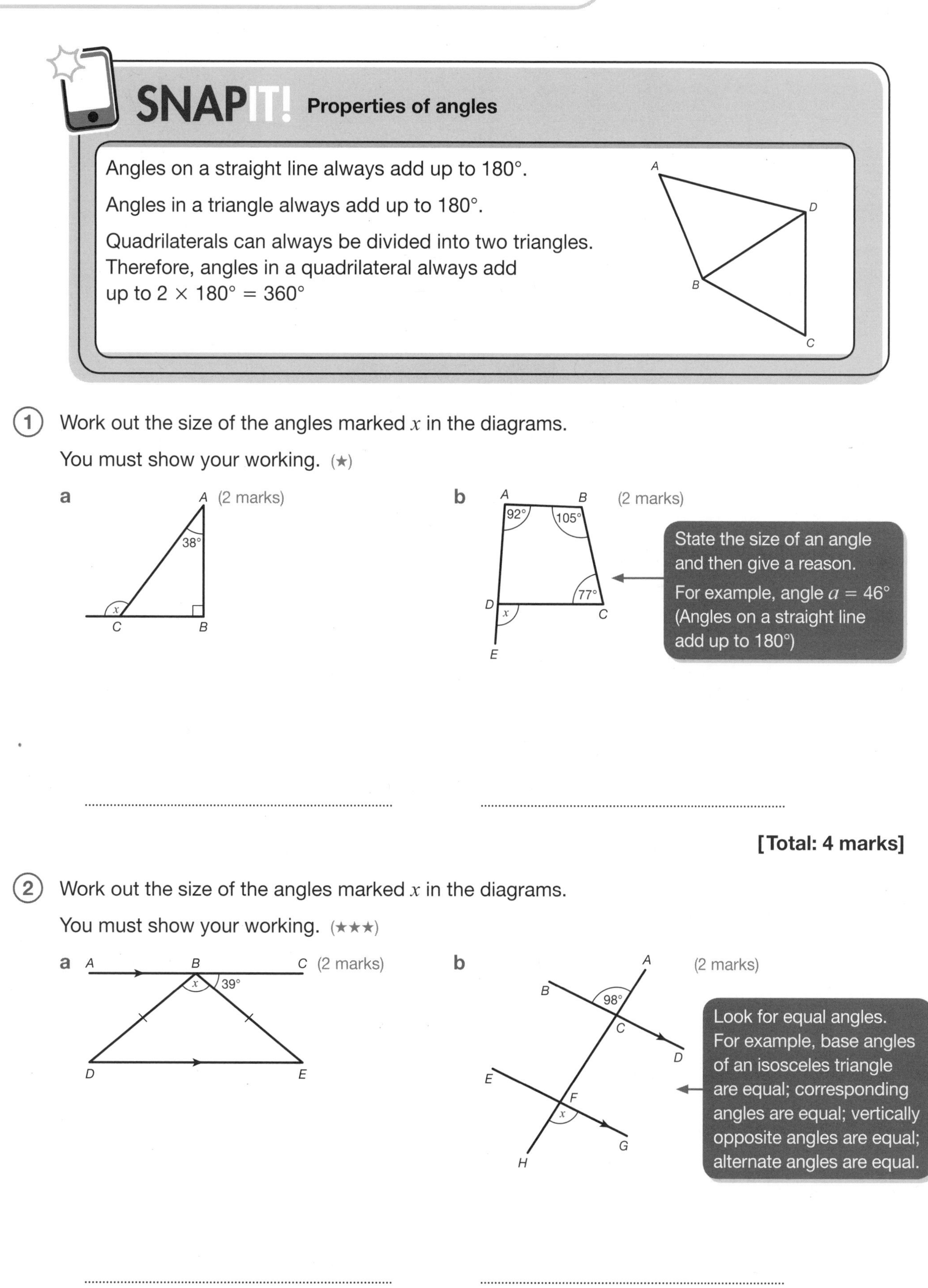

SNAP IT! Properties of angles

Angles on a straight line always add up to 180°.

Angles in a triangle always add up to 180°.

Quadrilaterals can always be divided into two triangles. Therefore, angles in a quadrilateral always add up to $2 \times 180° = 360°$

1. Work out the size of the angles marked x in the diagrams.

   You must show your working. (★)

   a (2 marks)

   b (2 marks)

   State the size of an angle and then give a reason.
   For example, angle $a = 46°$
   (Angles on a straight line add up to 180°)

   ....

   [Total: 4 marks]

2. Work out the size of the angles marked x in the diagrams.

   You must show your working. (★★★)

   a (2 marks)

   b (2 marks)

   Look for equal angles. For example, base angles of an isosceles triangle are equal; corresponding angles are equal; vertically opposite angles are equal; alternate angles are equal.

   ....

   [Total: 4 marks]

3. Work out the size of the angle marked x in the diagram.

   You must show your working. (2 marks, ★★★★)

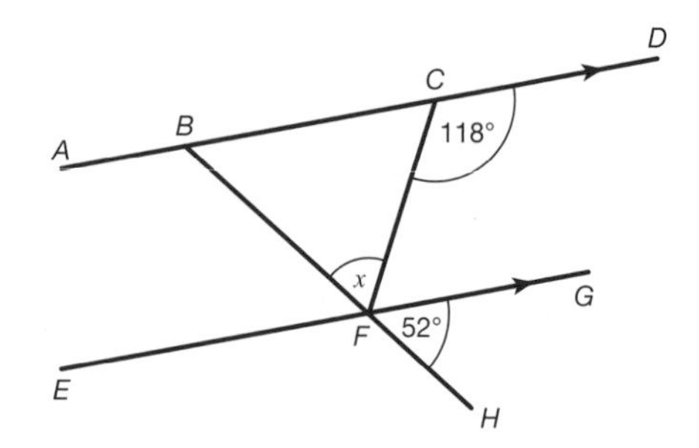

....

4. Show that this is an equilateral triangle. (3 marks, ★★★★★)

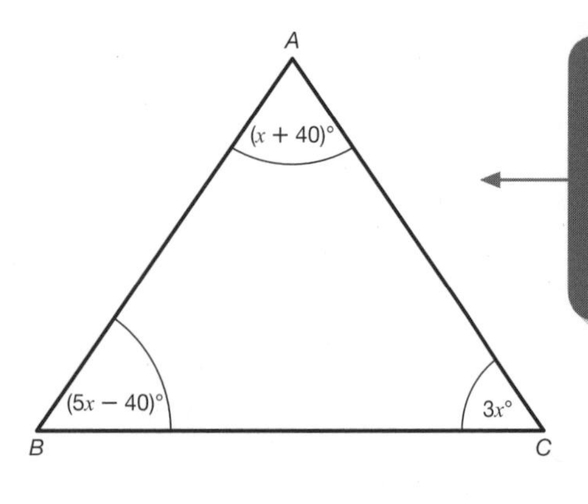

Write an equation and solve it to find x.

Then show that each angle is exactly as you would expect in an equilateral triangle.

Revise all the properties of angles. Now write every angle you can find on the diagram until you find x.

STRETCHIT!

A triangle has sides of 10, x and x cm. It has angles of 60°, θ° and θ°.

Can you say what kind of triangle it is?

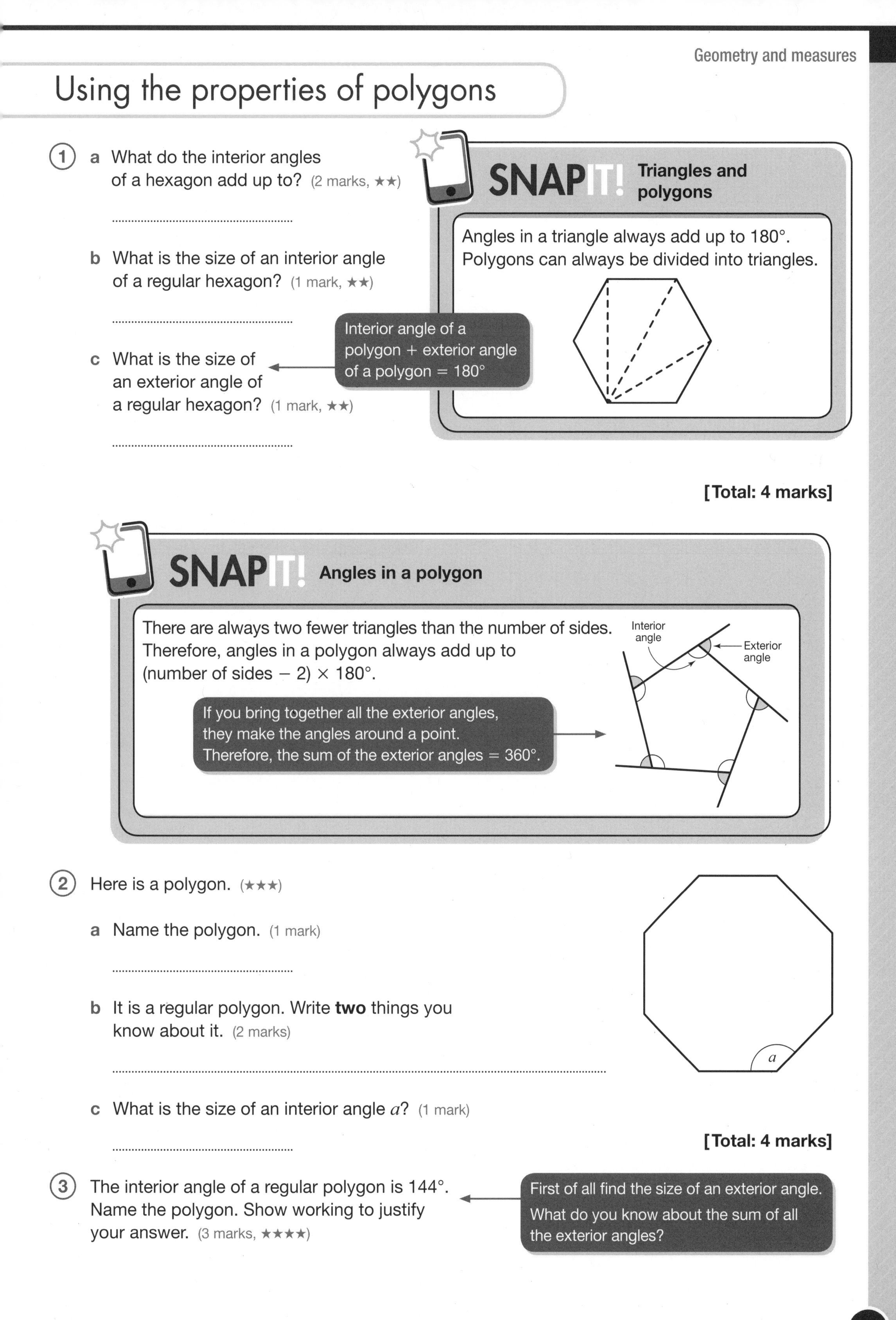

Using the properties of polygons

1. a What do the interior angles of a hexagon add up to? (2 marks, ★★)

   ....

   b What is the size of an interior angle of a regular hexagon? (1 mark, ★★)

   ....

   c What is the size of an exterior angle of a regular hexagon? (1 mark, ★★)

   ....

Interior angle of a polygon + exterior angle of a polygon = 180°

SNAP IT! Triangles and polygons

Angles in a triangle always add up to 180°.
Polygons can always be divided into triangles.

[Total: 4 marks]

SNAP IT! Angles in a polygon

There are always two fewer triangles than the number of sides. Therefore, angles in a polygon always add up to (number of sides − 2) × 180°.

If you bring together all the exterior angles, they make the angles around a point. Therefore, the sum of the exterior angles = 360°.

2. Here is a polygon. (★★★)

   a Name the polygon. (1 mark)

   ....

   b It is a regular polygon. Write **two** things you know about it. (2 marks)

   ....

   c What is the size of an interior angle a? (1 mark)

   ....

[Total: 4 marks]

3. The interior angle of a regular polygon is 144°. Name the polygon. Show working to justify your answer. (3 marks, ★★★★)

First of all find the size of an exterior angle. What do you know about the sum of all the exterior angles?

....

Using bearings

1 A ship sails from point A to point B to point C. (★★★)

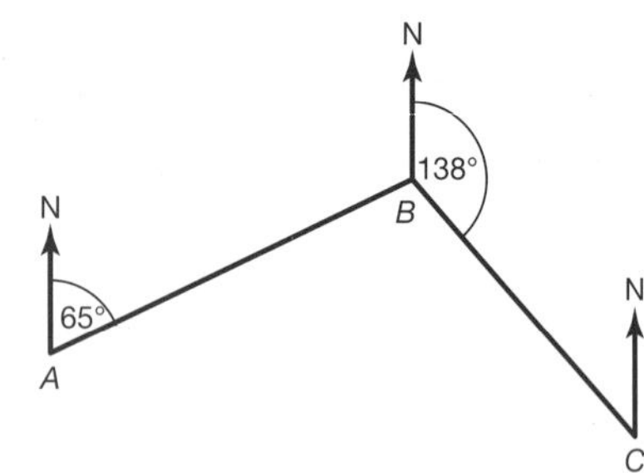

Bearings are always measured in degrees, clockwise from North. They are always written with three digits.

a What is the bearing of B from A? (1 mark)

....

b Work out the bearing of B from C. (1 mark)

The North arrows are parallel.
Therefore, use co-interior angles to find the angle between the line BC and the North arrow at C.
Then use angles at a point to find the bearing.

....

c Work out the bearing of A from B. (2 marks)

....

[Total: 4 marks]

2 The bearing of X from O is 276°. Work out the bearing of O from X. (2 marks, ★★★★)

Draw a diagram.

....

3 Here are points P and Q. (★★★★)

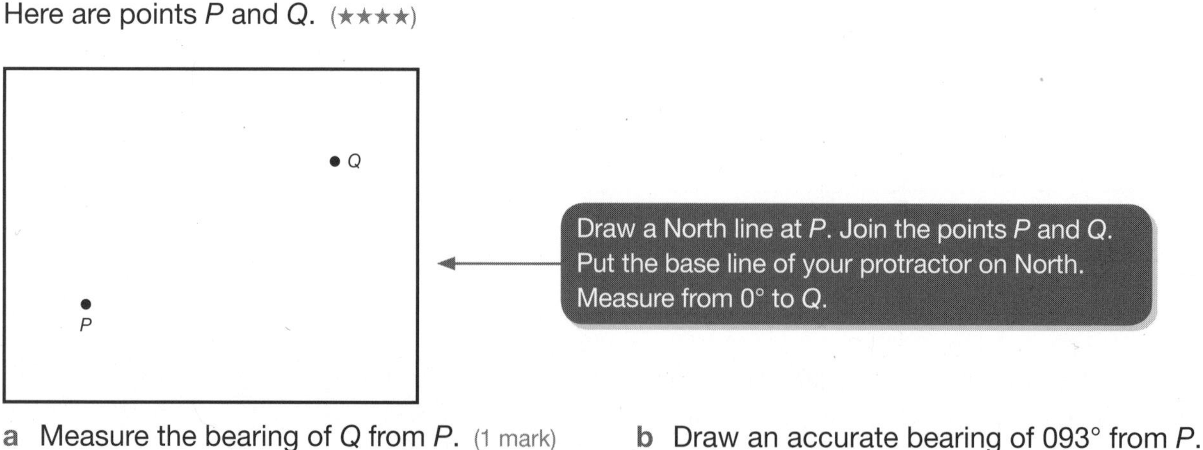

Draw a North line at P. Join the points P and Q. Put the base line of your protractor on North. Measure from 0° to Q.

a Measure the bearing of Q from P. (1 mark)

....

b Draw an accurate bearing of 093° from P.
Draw an accurate bearing of 225° from Q.
Mark the point where they intersect X. (2 marks)

[Total: 3 marks]

Properties of 2D shapes

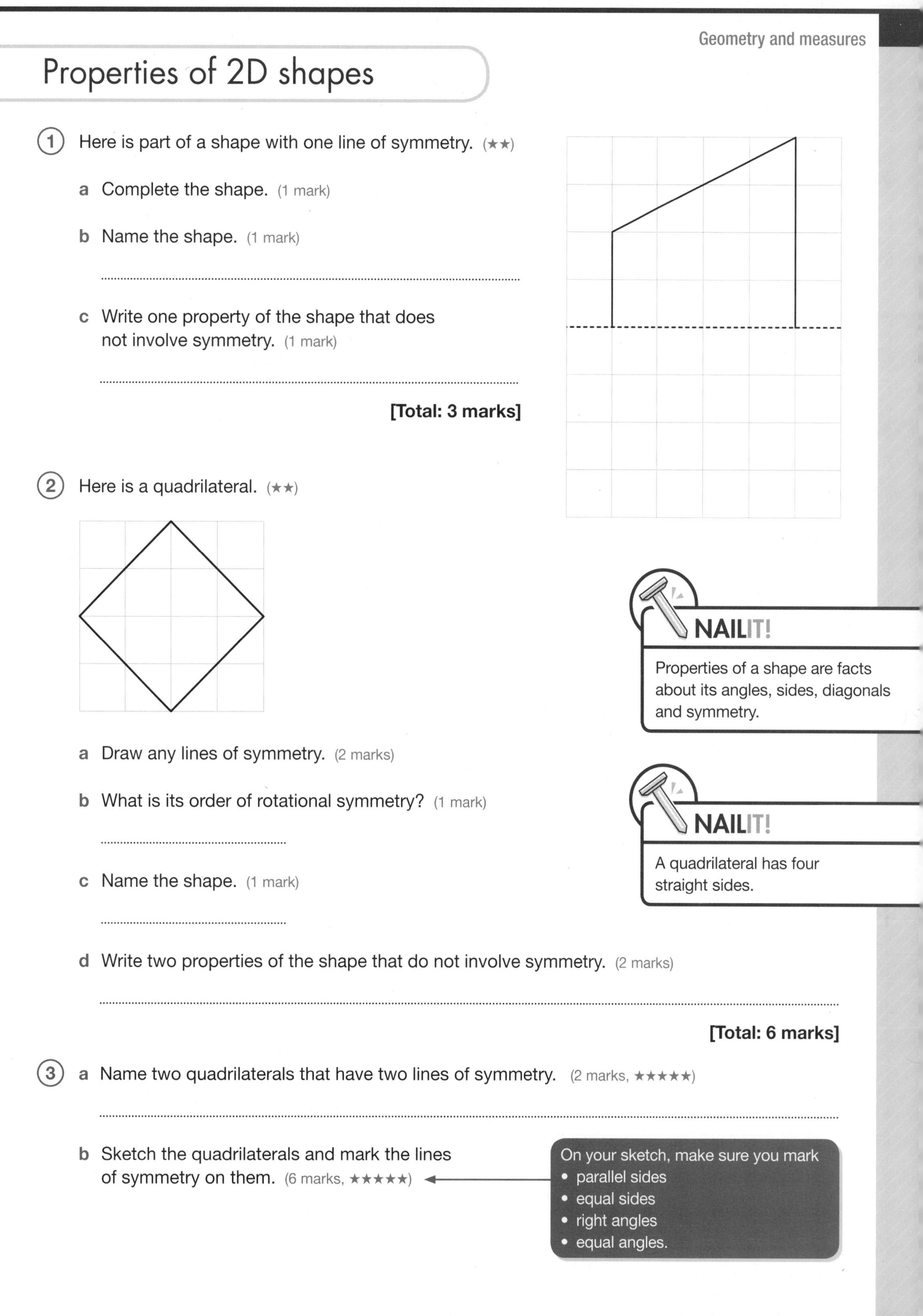

1. Here is part of a shape with one line of symmetry. (★★)

   a Complete the shape. (1 mark)

   b Name the shape. (1 mark)

   ....

   c Write one property of the shape that does not involve symmetry. (1 mark)

   ....

[Total: 3 marks]

2. Here is a quadrilateral. (★★)

   a Draw any lines of symmetry. (2 marks)

   b What is its order of rotational symmetry? (1 mark)

   ....

   c Name the shape. (1 mark)

   ....

   d Write two properties of the shape that do not involve symmetry. (2 marks)

   ....

[Total: 6 marks]

NAILIT!

Properties of a shape are facts about its angles, sides, diagonals and symmetry.

NAILIT!

A quadrilateral has four straight sides.

3. a Name two quadrilaterals that have two lines of symmetry. (2 marks, ★★★★★)

   ....

   b Sketch the quadrilaterals and mark the lines of symmetry on them. (6 marks, ★★★★★)

On your sketch, make sure you mark
- parallel sides
- equal sides
- right angles
- equal angles.

[Total: 8 marks]

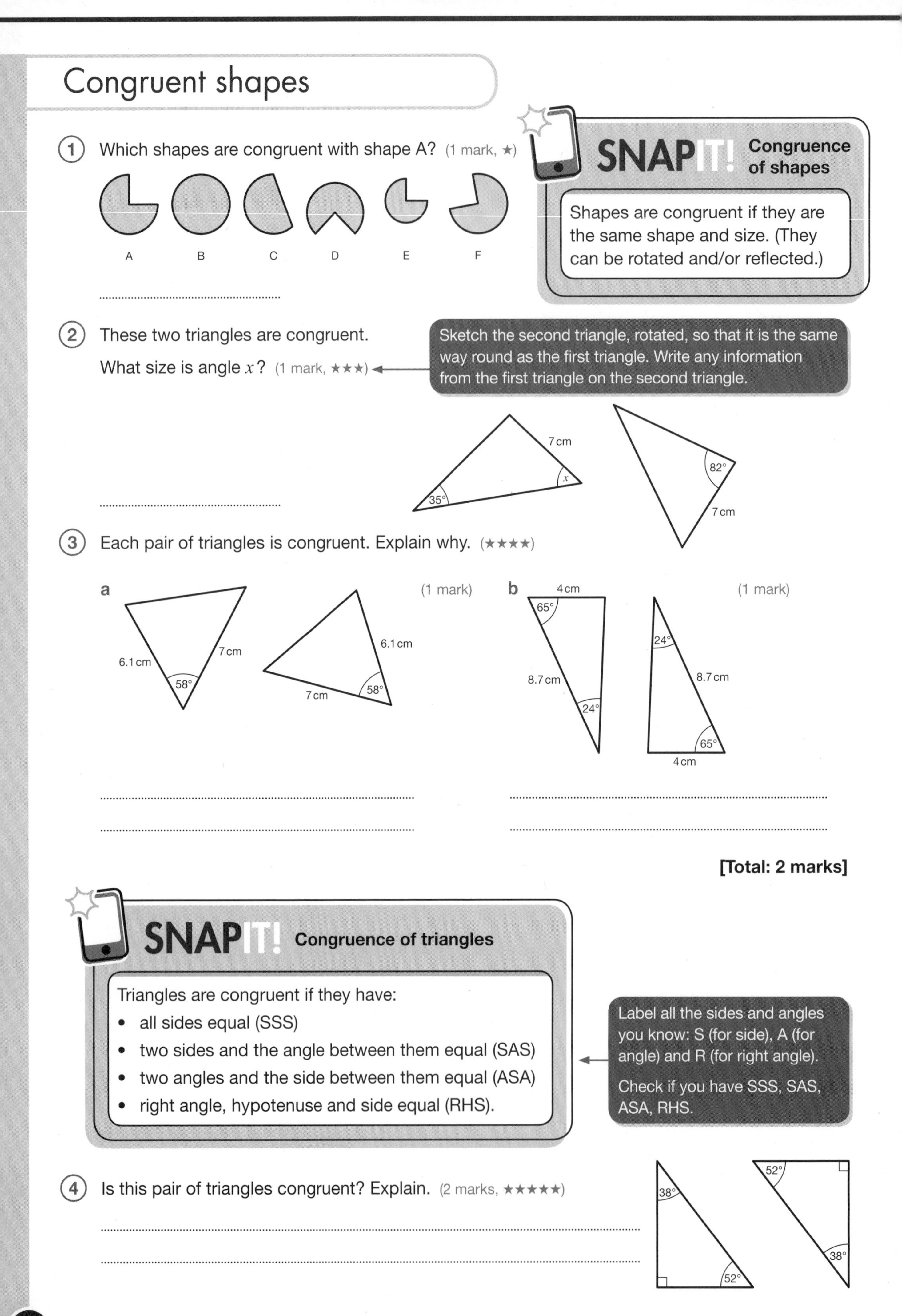

Congruent shapes

SNAPIT! Congruence of shapes

Shapes are congruent if they are the same shape and size. (They can be rotated and/or reflected.)

1. Which shapes are congruent with shape A? (1 mark, ★)

....

2. These two triangles are congruent.

   What size is angle x? (1 mark, ★★★)

   Sketch the second triangle, rotated, so that it is the same way round as the first triangle. Write any information from the first triangle on the second triangle.

....

3. Each pair of triangles is congruent. Explain why. (★★★★)

   a (1 mark)

   ....

   ....

   b (1 mark)

   ....

   ....

[Total: 2 marks]

SNAPIT! Congruence of triangles

Triangles are congruent if they have:

- all sides equal (SSS)
- two sides and the angle between them equal (SAS)
- two angles and the side between them equal (ASA)
- right angle, hypotenuse and side equal (RHS).

Label all the sides and angles you know: S (for side), A (for angle) and R (for right angle).

Check if you have SSS, SAS, ASA, RHS.

4. Is this pair of triangles congruent? Explain. (2 marks, ★★★★★)

....

....

Constructions

NAILIT!

When drawing constructions in maths, it is important to

- use a sharp pencil
- label any points, lengths or angles you are given
- *never* rub out your construction arcs and lines.

(1) **a** Use a ruler and compasses to construct an accurate drawing of triangle *ABC*. (3 marks, ★★★)

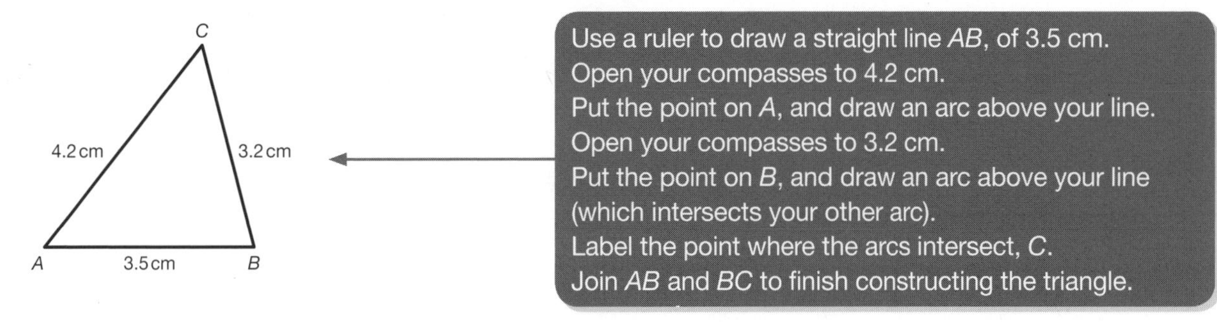

Use a ruler to draw a straight line *AB*, of 3.5 cm.
Open your compasses to 4.2 cm.
Put the point on *A*, and draw an arc above your line.
Open your compasses to 3.2 cm.
Put the point on *B*, and draw an arc above your line (which intersects your other arc).
Label the point where the arcs intersect, *C*.
Join *AB* and *BC* to finish constructing the triangle.

b Construct a bisector of angle *BAC*. (3 marks, ★★★)

The bisector cuts the angle exactly in half.
Open your compasses so that they are a little less than the lengths of the arms (*AC* or *AB*) of the angle *BAC*.
Keep them exactly like this throughout your construction.
Put the point of your compasses on *A*.
Draw two arcs – one on *AB* and one on *AC*.
Put the point of your compasses on the arc on *AB*.
Draw an arc in between the arms of the angle *BAC*.
Put the point of your compasses on the arc on *AC*.
Draw an arc in between the arms of the angle *BAC* (which intersects your other arc).
Join the point where the arcs intersect to *A*.

[Total: 6 marks]

(2) **a** Use a ruler and compasses to construct an accurate drawing of triangle *XYZ*, where $XY = 2.8$ cm, $XZ = 3.6$ cm, $YZ = 3$ cm. (3 marks, ★★★★)

Sketch and label the triangle first. Then you can see the triangle you are aiming to construct.

b Construct the bisector of angle *XYZ*. (3 marks, ★★★★)

[Total: 6 marks]

(3) **a** A person is standing at point O, in a field. He can see a straight road, PQ, ahead of him. (3 marks, ★★★★)

Use a ruler and compasses to construct the shortest distance from point O to the straight road.

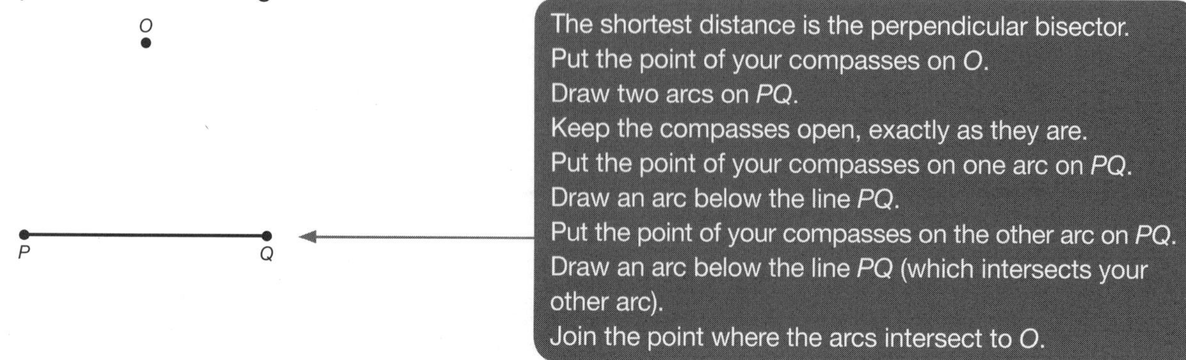

b The scale of this diagram is 1:100. Measure and write down the shortest distance. Give your answer in metres. (2 marks, ★★★★)

....

[Total: 5 marks]

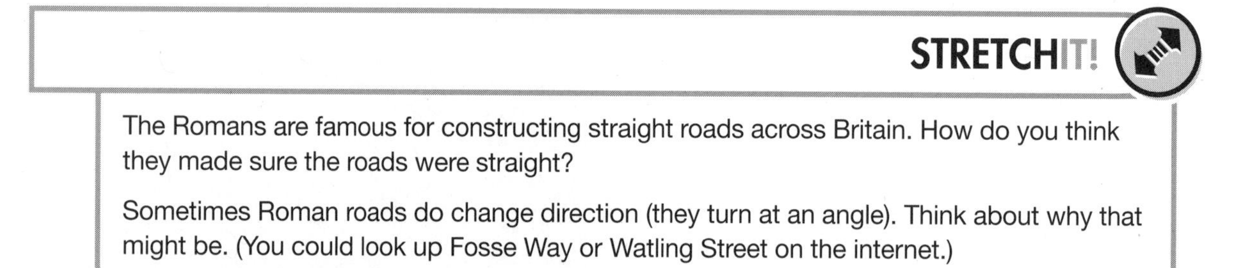

Drawing circles and parts of circles

1 a Draw a circle with diameter 7 cm. (1 mark, ★★)

You need to work out the radius to draw the circle.

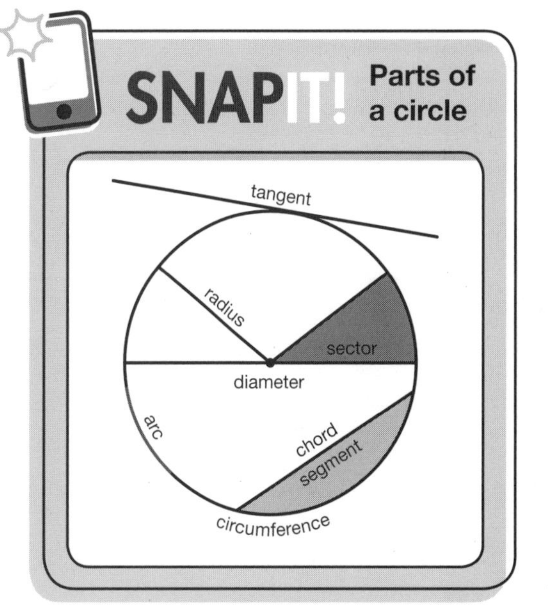

b Draw and label a tangent to the circle of length 4 cm. (1 mark, ★★)

c Shade and label a segment of the circle. (1 mark, ★★)

d Label a chord in the circle. (1 mark, ★★)

[Total: 4 marks]

2 a Draw a sector of a circle with radius 3 cm and angle 42°. (2 marks, ★★★)

b Label the arc. (1 mark, ★★★)

[Total: 3 marks]

3 Donald says, 'A segment of a circle and a sector of a circle are the same'.

Is Donald correct? You must give a reason for your answer. (3 marks, ★★★★)

....

....

....

Loci

① **a** Shade the locus of points less than 2.3 cm from point *A*. (2 marks, ★★★)

A

SNAPIT! Locus

A locus (plural: loci) is a set of points that obey a rule.

Draw the locus of points exactly 2.3 cm from the point. Shade the region less than 2.3 cm from the point.

b Draw the locus of points exactly 2.3 cm from line *BC*. (1 mark, ★★★)

B C

c Draw the locus of points equidistant from *DE* and *EF*. (2 marks, ★★★)

Equidistant means the same distance.

SNAPIT! Bisector

The locus of points equidistant from two lines that meet to make an angle is the bisector of the angle.

D

E

F

Don't forget to leave in any construction arcs and lines you use when drawing loci.

[Total: 5 marks]

(2) A path is laid so that it is exactly the same distance between a tree and a barn.

Draw the path. (2 marks, ★★★)

Barn

Tree

Imagine two points, A and B, joined by a line. The locus of points exactly the same distance from A and B is the perpendicular bisector of AB.

(3) This square field *ABCD* is drawn to scale: 1:200

1:200 means 1 cm on the scale diagram represents 200 cm = 2 m in real life

A dog and a goat are tethered in the field. The dog is tethered at *A* with a 4 m length of rope. The goat is tethered to a piece of wood nailed between two poles, at *X* and *Y*, with a 3 m length of rope. The rope can slide along the wood, between the poles.

Complete the scale drawing and shade the area that neither the dog, nor the goat, can reach. (3 marks, ★★★★)

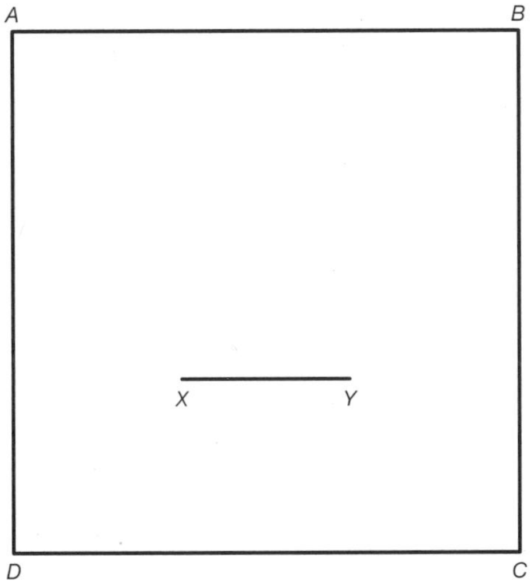

A *B*

X *Y*

D *C*

(4) Two cows graze in a field of length 20 m and width 12 m. They are tethered to diagonally opposite corners by ropes of length 13 m. On a scale drawing, show the area grazed by **both** cows. (3 marks, ★★★★)

Perimeter

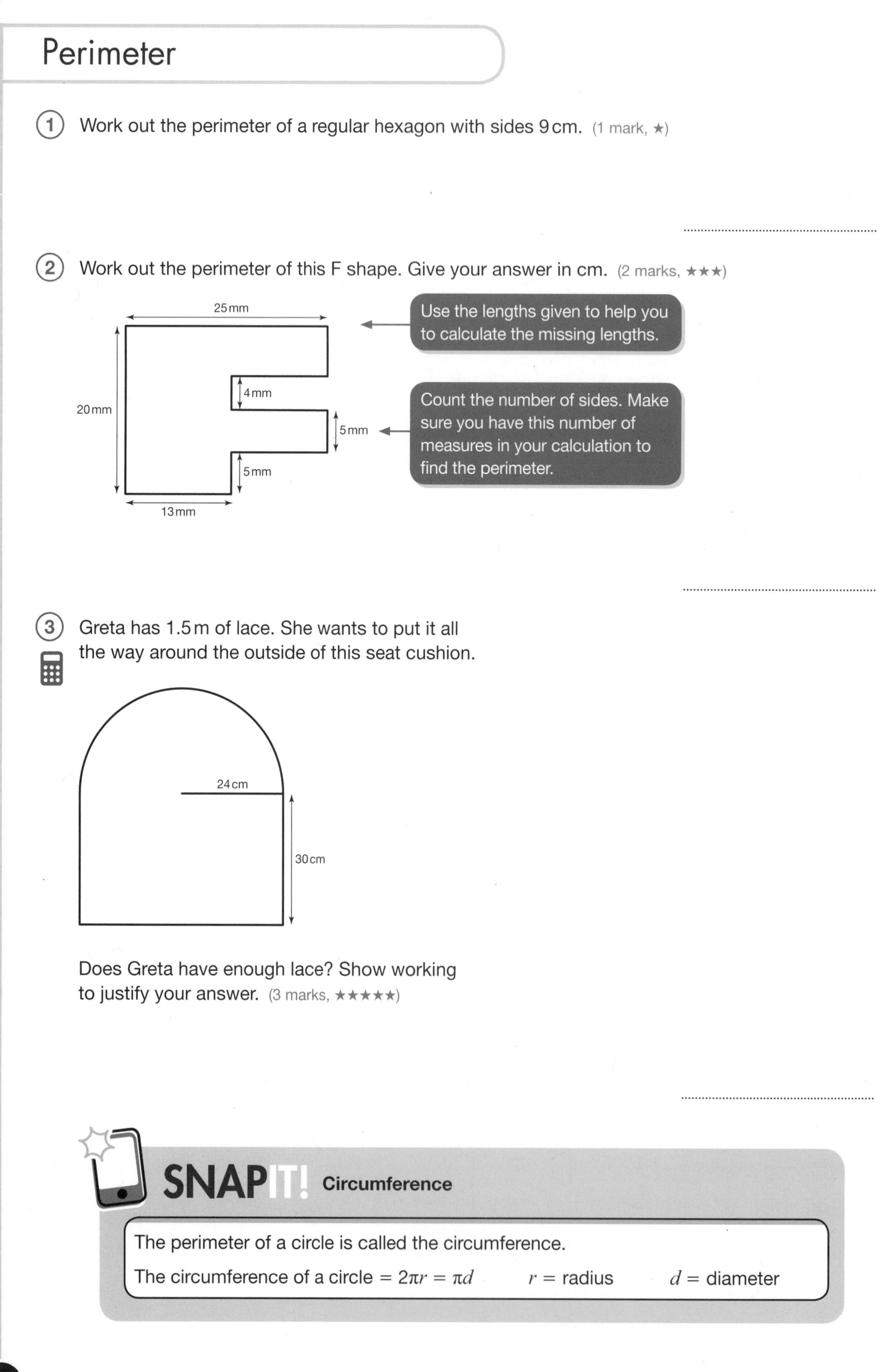

1. Work out the perimeter of a regular hexagon with sides 9 cm. (1 mark, ★)

....

2. Work out the perimeter of this F shape. Give your answer in cm. (2 marks, ★★★)

25 mm, 20 mm, 4 mm, 5 mm, 5 mm, 13 mm

Use the lengths given to help you to calculate the missing lengths.

Count the number of sides. Make sure you have this number of measures in your calculation to find the perimeter.

....

3. Greta has 1.5 m of lace. She wants to put it all the way around the outside of this seat cushion.

24 cm, 30 cm

Does Greta have enough lace? Show working to justify your answer. (3 marks, ★★★★★)

....

SNAP IT! Circumference

The perimeter of a circle is called the circumference.

The circumference of a circle $= 2\pi r = \pi d$ $\quad r =$ radius $\quad d =$ diameter

Area

(1) What are the areas of these shapes? (★★)

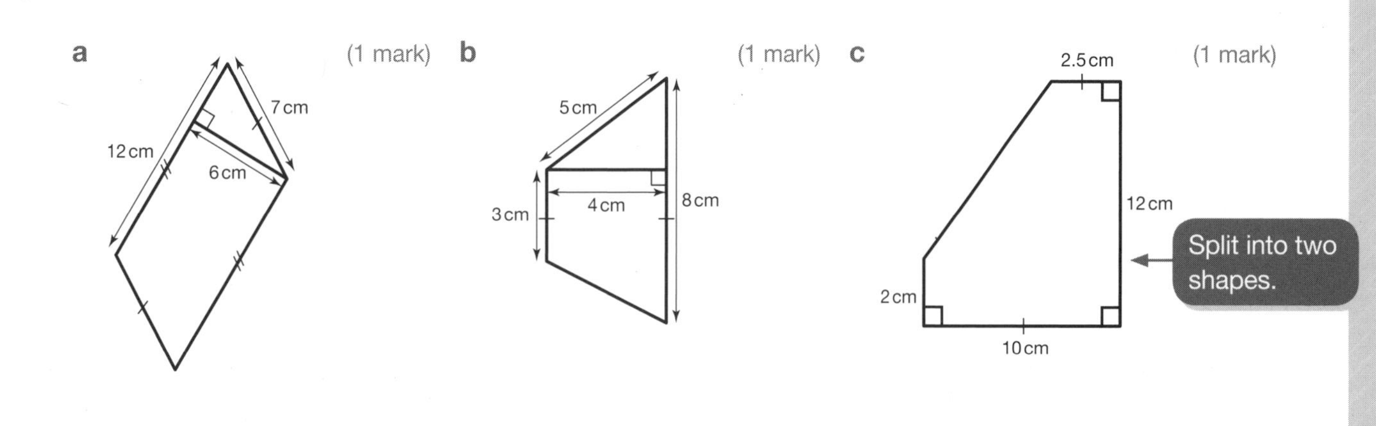

....

[Total: 3 marks]

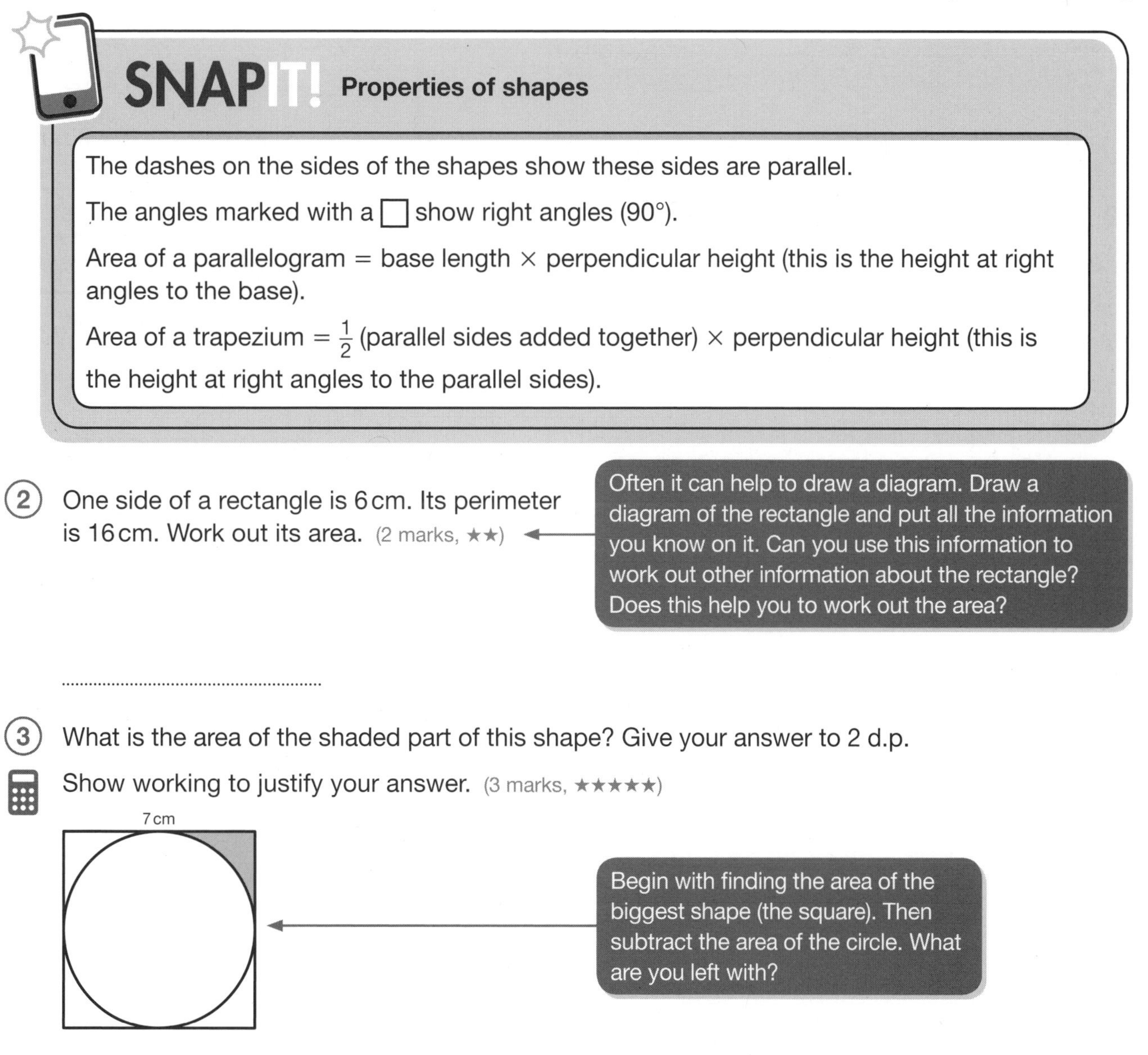

SNAP IT! Properties of shapes

The dashes on the sides of the shapes show these sides are parallel.

The angles marked with a ☐ show right angles (90°).

Area of a parallelogram = base length × perpendicular height (this is the height at right angles to the base).

Area of a trapezium = $\frac{1}{2}$ (parallel sides added together) × perpendicular height (this is the height at right angles to the parallel sides).

(2) One side of a rectangle is 6 cm. Its perimeter is 16 cm. Work out its area. (2 marks, ★★)

Often it can help to draw a diagram. Draw a diagram of the rectangle and put all the information you know on it. Can you use this information to work out other information about the rectangle? Does this help you to work out the area?

....

(3) What is the area of the shaded part of this shape? Give your answer to 2 d.p.

Show working to justify your answer. (3 marks, ★★★★★)

7 cm

Begin with finding the area of the biggest shape (the square). Then subtract the area of the circle. What are you left with?

....

Sectors

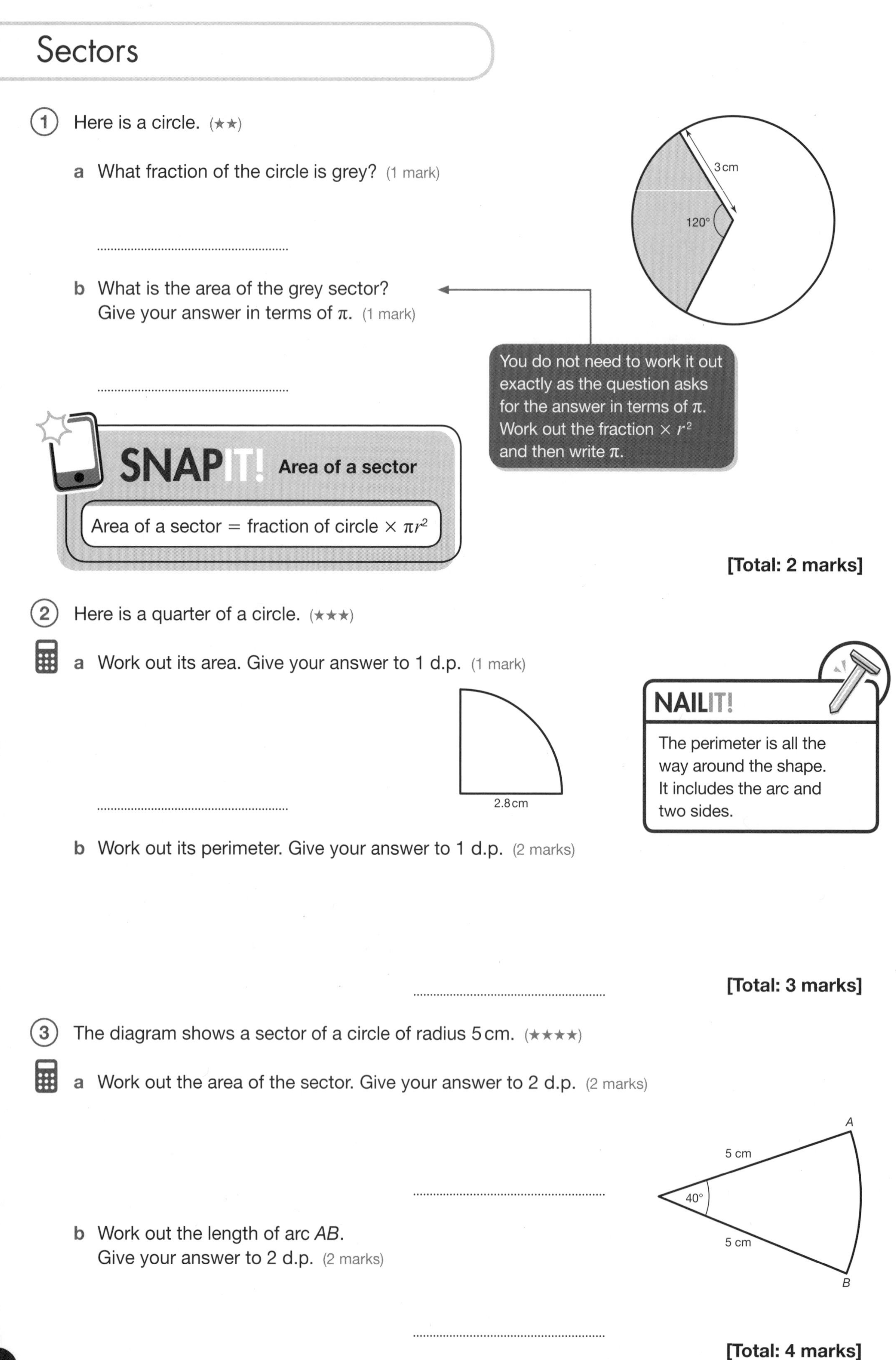

1. Here is a circle. (★★)

   a What fraction of the circle is grey? (1 mark)

   ....

   b What is the area of the grey sector?
   Give your answer in terms of π. (1 mark)

   ....

You do not need to work it out exactly as the question asks for the answer in terms of π. Work out the fraction $\times\ r^2$ and then write π.

SNAPIT! **Area of a sector**

Area of a sector = fraction of circle $\times\ \pi r^2$

[Total: 2 marks]

2. Here is a quarter of a circle. (★★★)

   a Work out its area. Give your answer to 1 d.p. (1 mark)

   ....

   b Work out its perimeter. Give your answer to 1 d.p. (2 marks)

   ....

NAILIT!

The perimeter is all the way around the shape. It includes the arc and two sides.

[Total: 3 marks]

3. The diagram shows a sector of a circle of radius 5 cm. (★★★★)

   a Work out the area of the sector. Give your answer to 2 d.p. (2 marks)

   ....

   b Work out the length of arc *AB*.
   Give your answer to 2 d.p. (2 marks)

   ....

[Total: 4 marks]

3D shapes

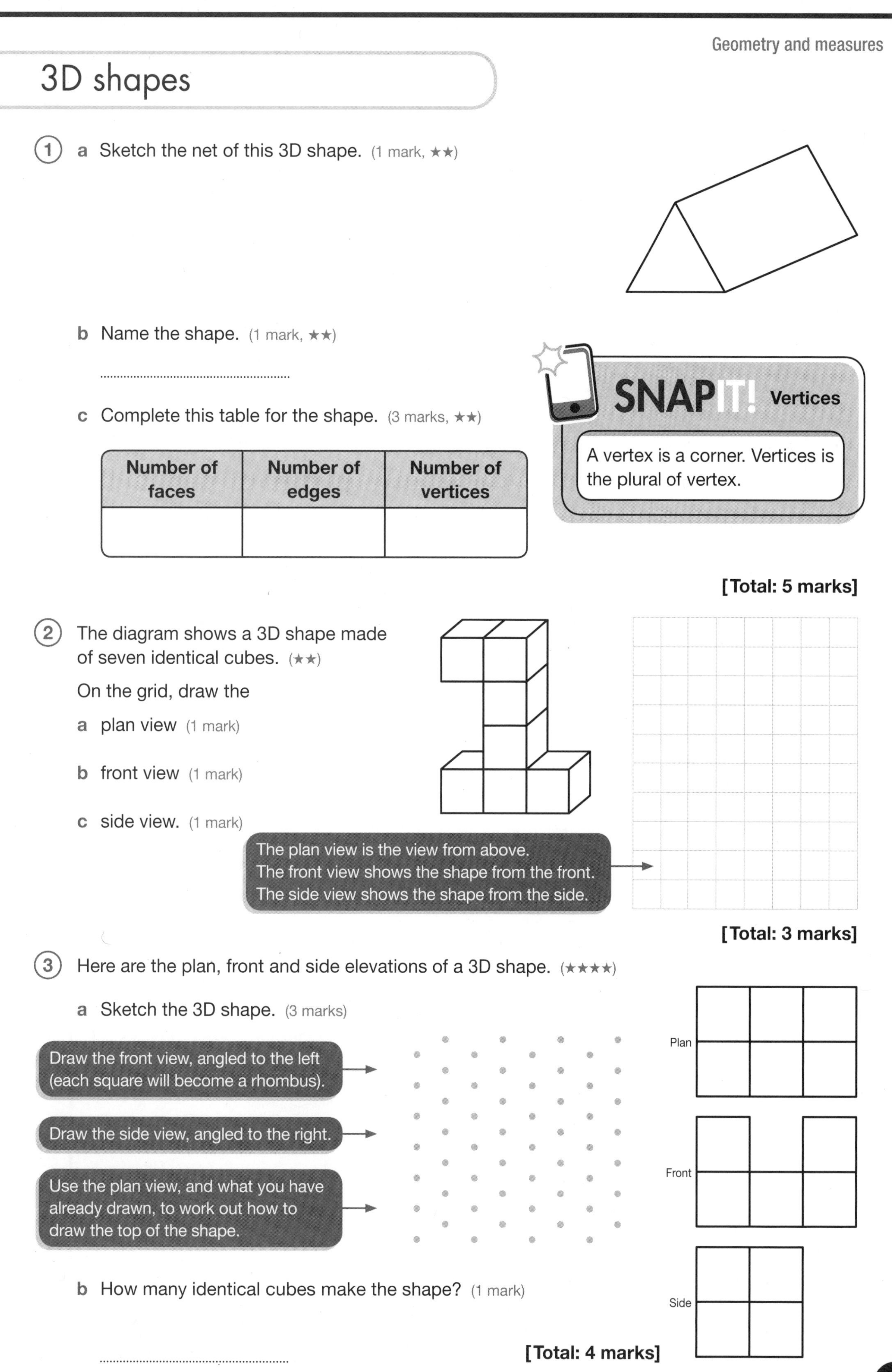

1 a Sketch the net of this 3D shape. (1 mark, ★★)

b Name the shape. (1 mark, ★★)

....

c Complete this table for the shape. (3 marks, ★★)

Number of faces	Number of edges	Number of vertices

SNAP IT! Vertices

A vertex is a corner. Vertices is the plural of vertex.

[Total: 5 marks]

2 The diagram shows a 3D shape made of seven identical cubes. (★★)

On the grid, draw the

a plan view (1 mark)

b front view (1 mark)

c side view. (1 mark)

The plan view is the view from above.
The front view shows the shape from the front.
The side view shows the shape from the side.

[Total: 3 marks]

3 Here are the plan, front and side elevations of a 3D shape. (★★★★)

a Sketch the 3D shape. (3 marks)

Draw the front view, angled to the left (each square will become a rhombus).

Draw the side view, angled to the right.

Use the plan view, and what you have already drawn, to work out how to draw the top of the shape.

b How many identical cubes make the shape? (1 mark)

....

[Total: 4 marks]

Volume

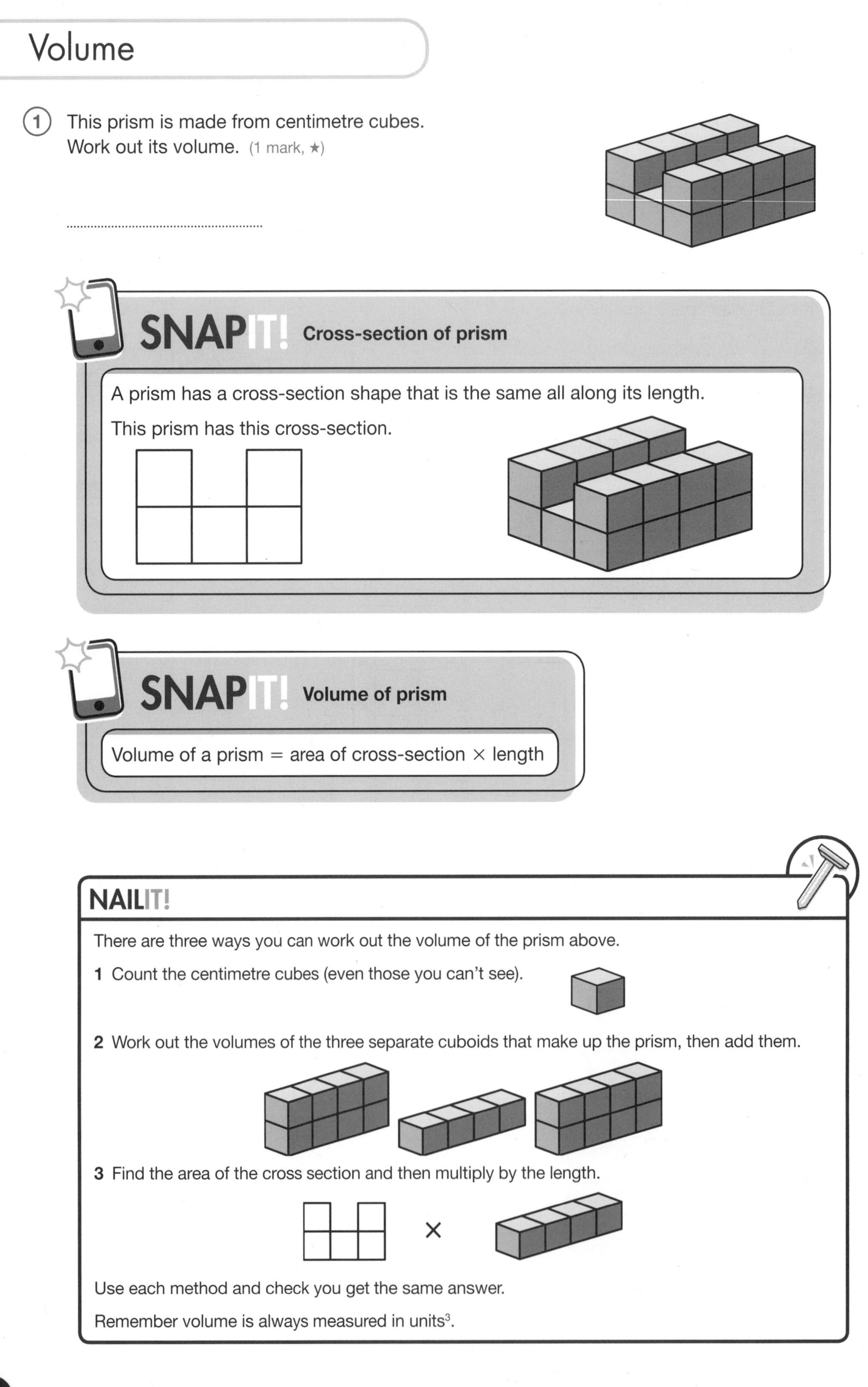

1. This prism is made from centimetre cubes.
   Work out its volume. (1 mark, ★)

   ....

SNAPIT! Cross-section of prism

A prism has a cross-section shape that is the same all along its length.

This prism has this cross-section.

SNAPIT! Volume of prism

Volume of a prism = area of cross-section × length

NAILIT!

There are three ways you can work out the volume of the prism above.

1 Count the centimetre cubes (even those you can't see).

2 Work out the volumes of the three separate cuboids that make up the prism, then add them.

3 Find the area of the cross section and then multiply by the length.

×

Use each method and check you get the same answer.

Remember volume is always measured in units^3.

(2) Work out the volume of these shapes. Give your answer to the nearest cm^3. (★★★)

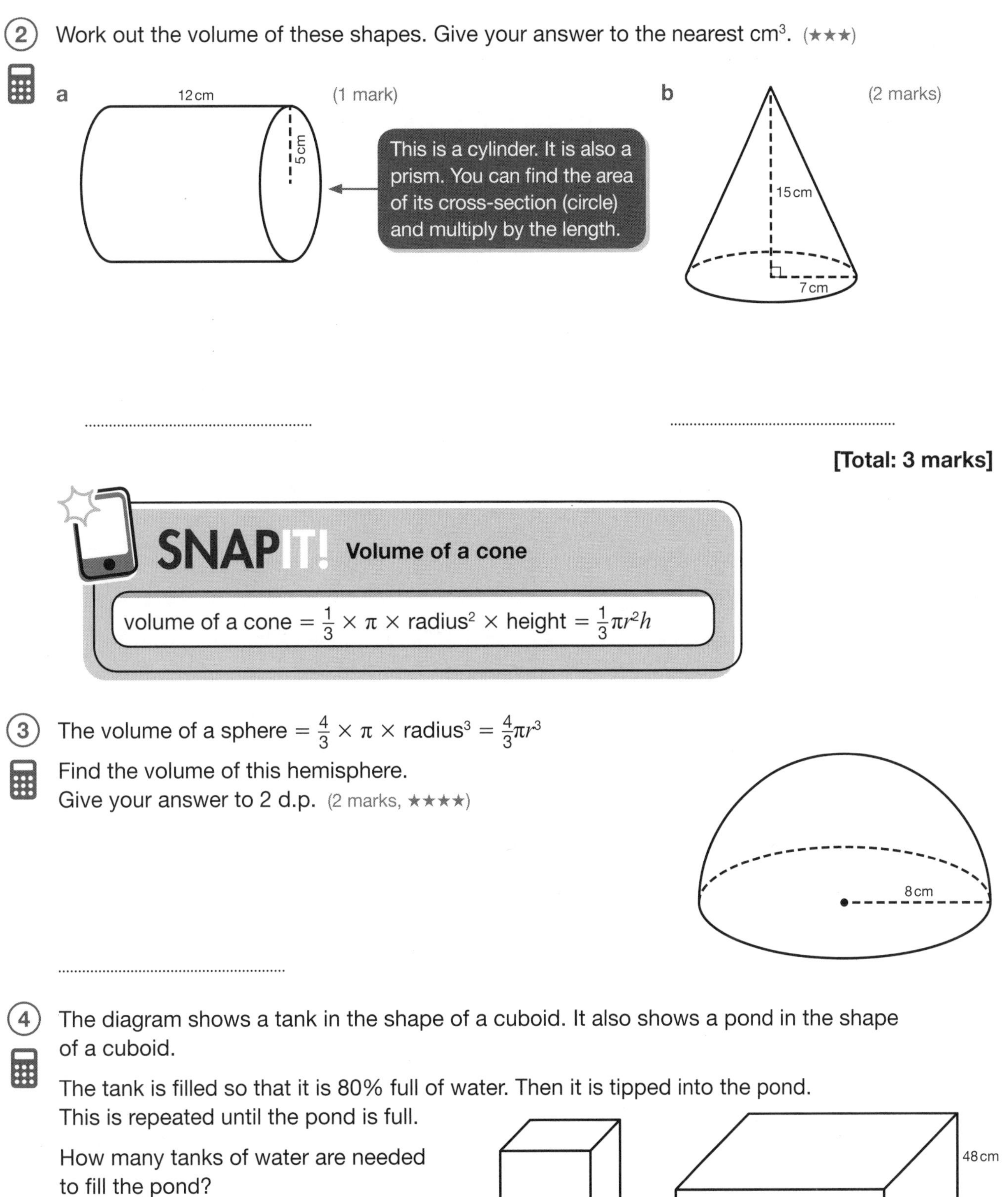

....

[Total: 3 marks]

SNAPIT! Volume of a cone

volume of a cone $= \frac{1}{3} \times \pi \times \text{radius}^2 \times \text{height} = \frac{1}{3}\pi r^2 h$

(3) The volume of a sphere $= \frac{4}{3} \times \pi \times \text{radius}^3 = \frac{4}{3}\pi r^3$

Find the volume of this hemisphere.
Give your answer to 2 d.p. (2 marks, ★★★★)

....

(4) The diagram shows a tank in the shape of a cuboid. It also shows a pond in the shape of a cuboid.

The tank is filled so that it is 80% full of water. Then it is tipped into the pond. This is repeated until the pond is full.

How many tanks of water are needed to fill the pond?

Show working to justify your answer. (4 marks, ★★★★★)

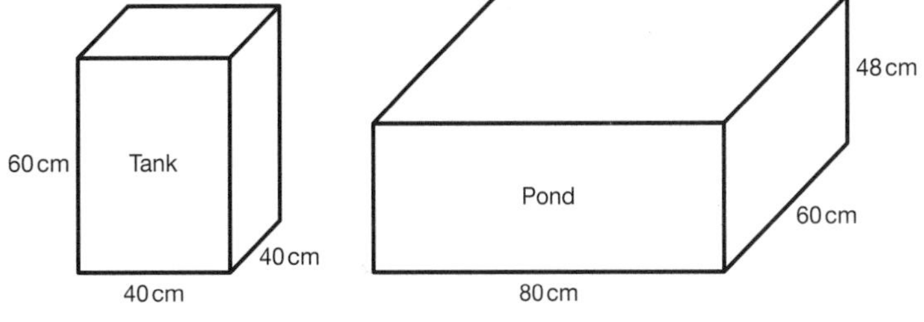

....

Surface area

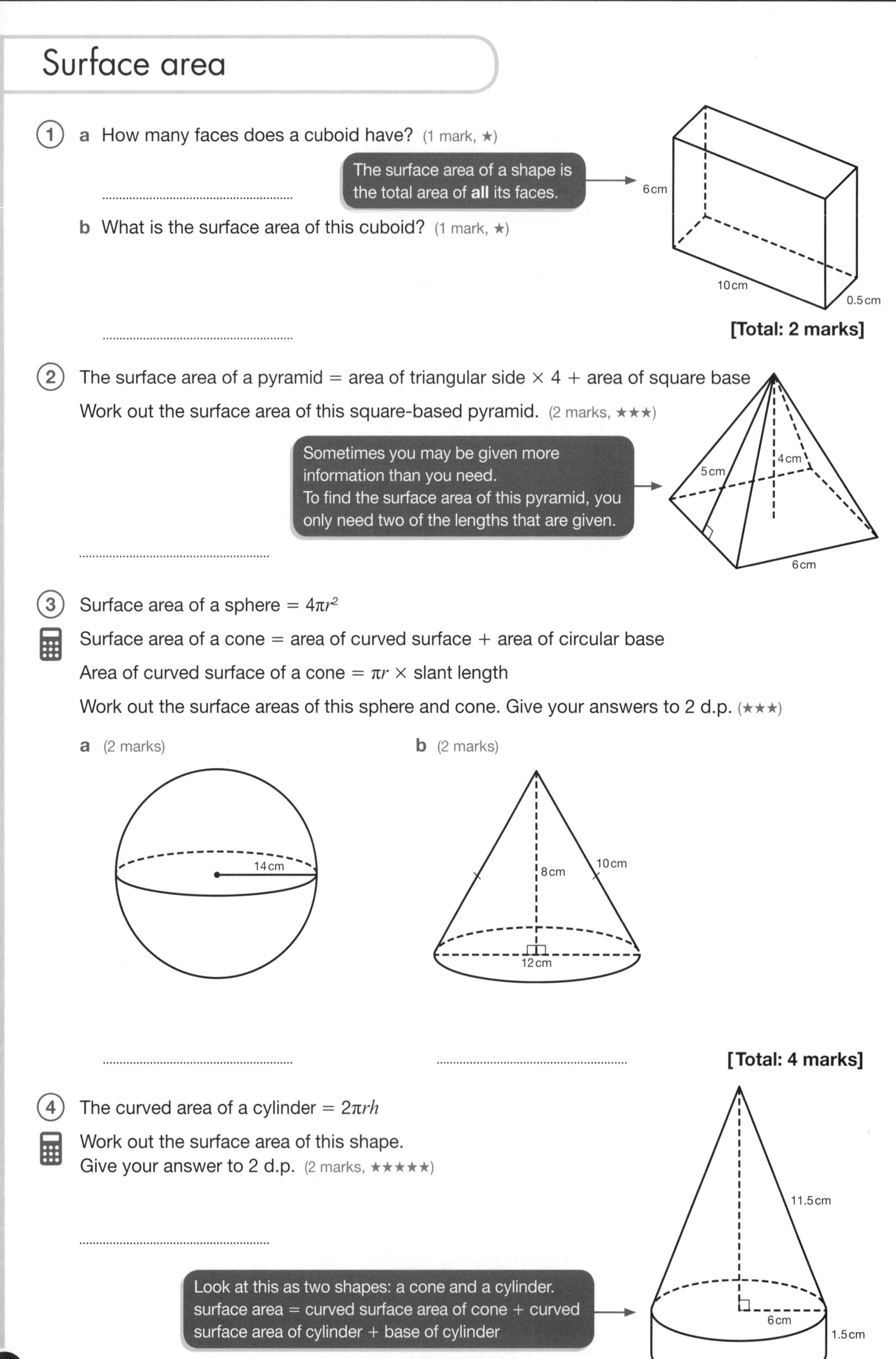

① **a** How many faces does a cuboid have? (1 mark, ★)

The surface area of a shape is the total area of **all** its faces.

....

b What is the surface area of this cuboid? (1 mark, ★)

....

[Total: 2 marks]

② The surface area of a pyramid = area of triangular side × 4 + area of square base

Work out the surface area of this square-based pyramid. (2 marks, ★★★)

Sometimes you may be given more information than you need.
To find the surface area of this pyramid, you only need two of the lengths that are given.

....

③ Surface area of a sphere = $4\pi r^2$

Surface area of a cone = area of curved surface + area of circular base

Area of curved surface of a cone = πr × slant length

Work out the surface areas of this sphere and cone. Give your answers to 2 d.p. (★★★)

a (2 marks)

b (2 marks)

....

[Total: 4 marks]

④ The curved area of a cylinder = $2\pi rh$

Work out the surface area of this shape.
Give your answer to 2 d.p. (2 marks, ★★★★★)

....

Look at this as two shapes: a cone and a cylinder.
surface area = curved surface area of cone + curved surface area of cylinder + base of cylinder

Using Pythagoras' theorem

SNAPIT! Pythagoras' theorem

For a right-angled triangle

$c^2 = a^2 + b^2$

where c is the hypotenuse (the longest side, opposite the right angle) and a and b are the two shorter sides.

1 Work out the missing side in these two triangles. (2 marks, ★★)

4cm
3cm
x
12cm
y
15cm

....

DOIT!

For each triangle.

1 Identify the hypotenuse. Label this c.

2 Substitute the lengths you know into $c^2 = a^2 + b^2$.

3 Don't forget to square root to find the missing side.

2 A rectangular lawn is 4.5 m long. This is 1.5 m shorter than its diagonal. Work out the area of the rectangle. Give your answer to 1 d.p. (3 marks, ★★★)

Read the first sentence. Draw a diagram. Read the next sentence. Add the information to your diagram. Underline what the question is asking – the area.

....

3 Point A has coordinates (2, 1); point B has coordinates (−2, −1). Work out the length of the line AB. Give your answer in surd form. (3 marks, ★★★★)

Sketch the points on axes. Draw a right angled triangle so that the line AB is the hypotenuse. Now use Pythagoras' theorem. Your answer should be in the form ■√■

....

4 A doorway is 70 cm wide and 190 cm high. A large piece of artwork is 2 m × 2 m. Will it fit through the doorway? Show working to justify your answer. (3 marks, ★★★★★)

....

Trigonometry

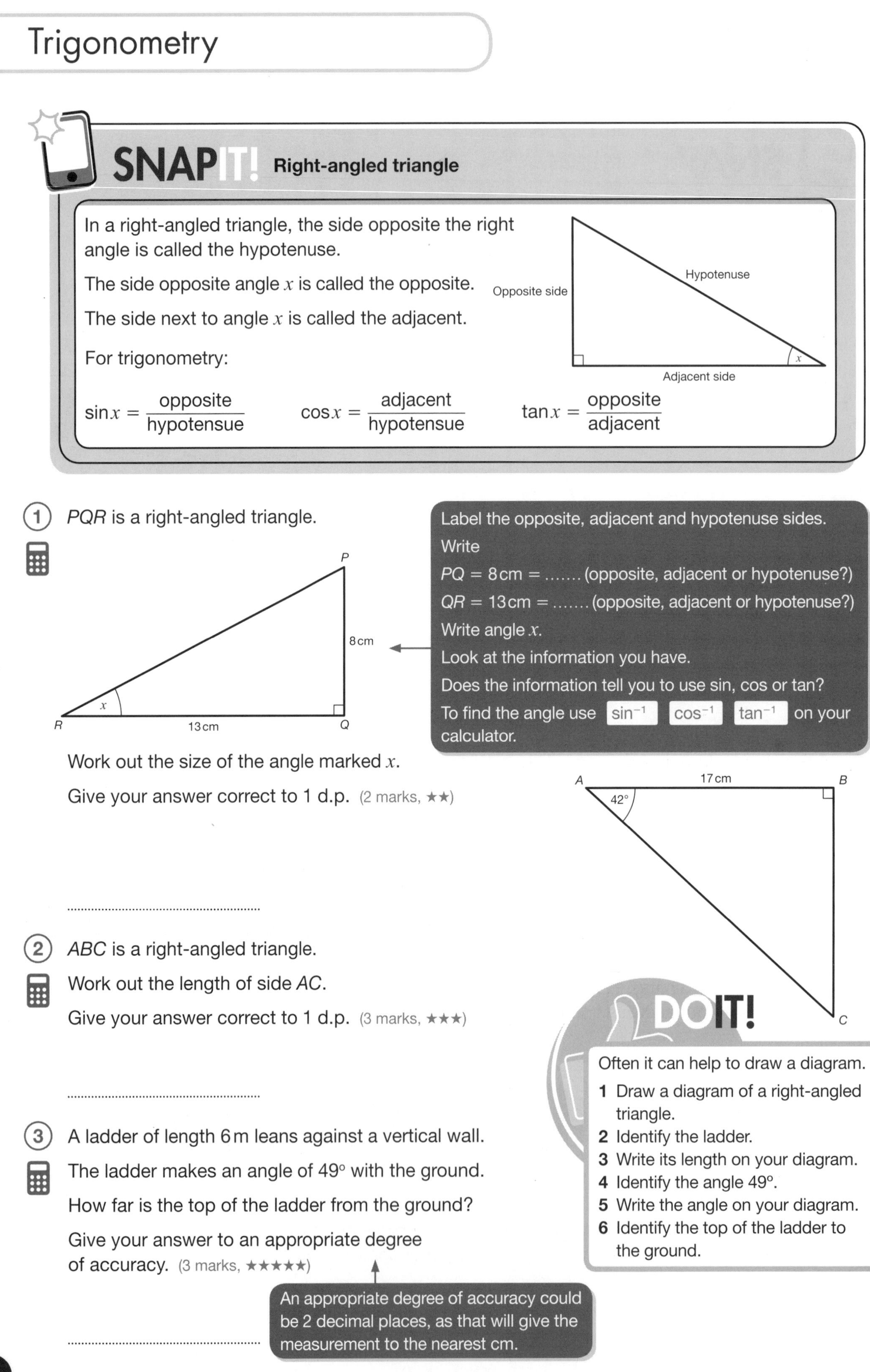

SNAPIT! Right-angled triangle

In a right-angled triangle, the side opposite the right angle is called the hypotenuse.

The side opposite angle x is called the opposite.

The side next to angle x is called the adjacent.

For trigonometry:

$\sin x = \dfrac{\text{opposite}}{\text{hypotenuse}}$ $\quad \cos x = \dfrac{\text{adjacent}}{\text{hypotenuse}}$ $\quad \tan x = \dfrac{\text{opposite}}{\text{adjacent}}$

1. *PQR* is a right-angled triangle.

Label the opposite, adjacent and hypotenuse sides.
Write
$PQ = 8\text{cm} = \ldots\ldots\ldots$ (opposite, adjacent or hypotenuse?)
$QR = 13\text{cm} = \ldots\ldots\ldots$ (opposite, adjacent or hypotenuse?)
Write angle x.
Look at the information you have.
Does the information tell you to use sin, cos or tan?
To find the angle use $\sin^{-1}$ $\cos^{-1}$ $\tan^{-1}$ on your calculator.

Work out the size of the angle marked x.

Give your answer correct to 1 d.p. (2 marks, ★★)

....

2. *ABC* is a right-angled triangle.

Work out the length of side *AC*.

Give your answer correct to 1 d.p. (3 marks, ★★★)

....

3. A ladder of length 6m leans against a vertical wall.

The ladder makes an angle of 49° with the ground.

How far is the top of the ladder from the ground?

Give your answer to an appropriate degree of accuracy. (3 marks, ★★★★★)

An appropriate degree of accuracy could be 2 decimal places, as that will give the measurement to the nearest cm.

....

DOIT!

Often it can help to draw a diagram.

1. Draw a diagram of a right-angled triangle.
2. Identify the ladder.
3. Write its length on your diagram.
4. Identify the angle 49°.
5. Write the angle on your diagram.
6. Identify the top of the ladder to the ground.

Exact trigonometric values

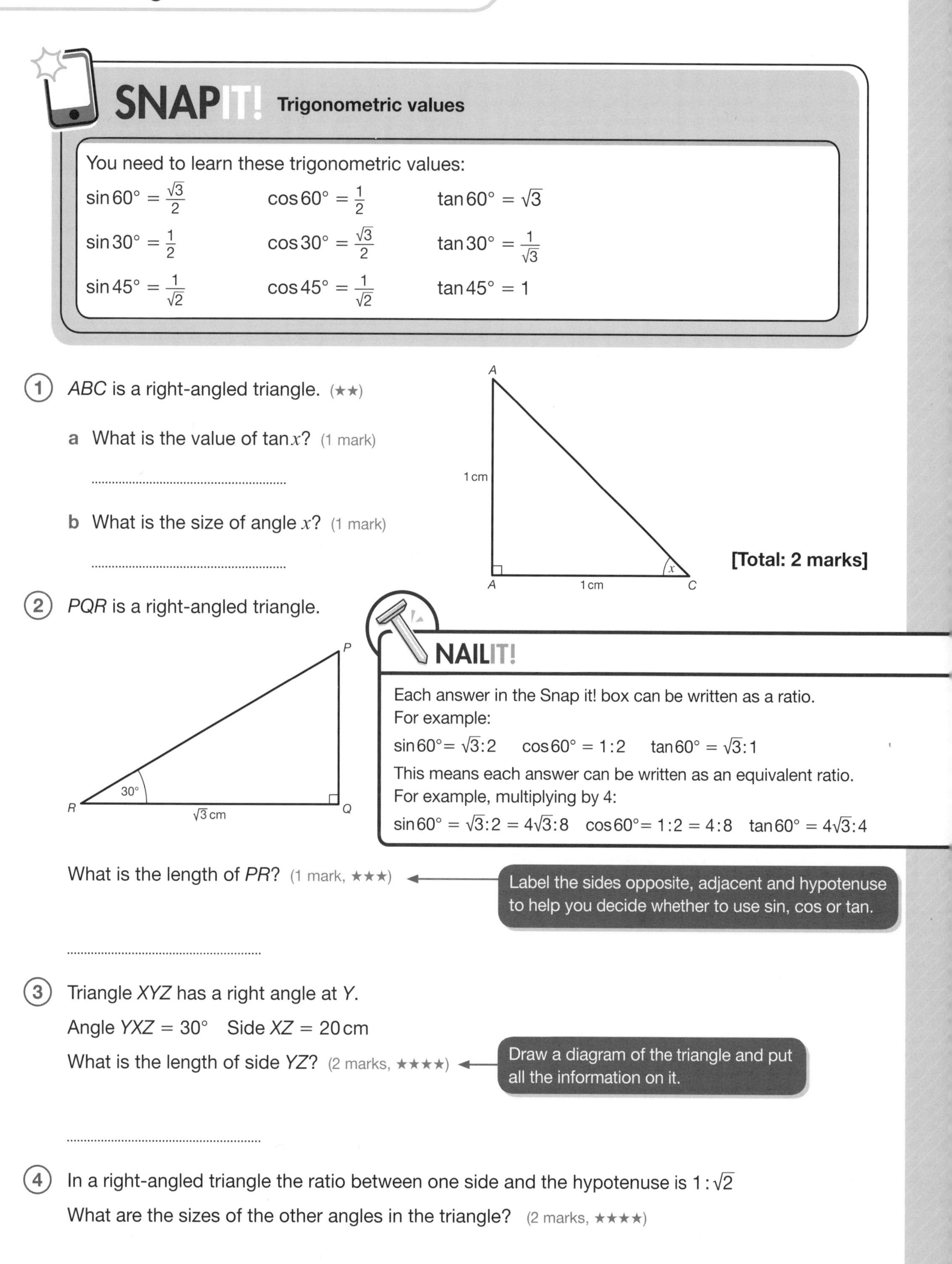

SNAP IT! Trigonometric values

You need to learn these trigonometric values:

$\sin 60° = \frac{\sqrt{3}}{2}$ $\quad$ $\cos 60° = \frac{1}{2}$ $\quad$ $\tan 60° = \sqrt{3}$

$\sin 30° = \frac{1}{2}$ $\quad$ $\cos 30° = \frac{\sqrt{3}}{2}$ $\quad$ $\tan 30° = \frac{1}{\sqrt{3}}$

$\sin 45° = \frac{1}{\sqrt{2}}$ $\quad$ $\cos 45° = \frac{1}{\sqrt{2}}$ $\quad$ $\tan 45° = 1$

1. *ABC* is a right-angled triangle. (★★)

   a What is the value of $\tan x$? (1 mark)

   ....

   b What is the size of angle x? (1 mark)

   ....

[Total: 2 marks]

2. *PQR* is a right-angled triangle.

NAIL IT!

Each answer in the Snap it! box can be written as a ratio.
For example:

$\sin 60° = \sqrt{3}:2$ $\quad$ $\cos 60° = 1:2$ $\quad$ $\tan 60° = \sqrt{3}:1$

This means each answer can be written as an equivalent ratio.
For example, multiplying by 4:

$\sin 60° = \sqrt{3}:2 = 4\sqrt{3}:8$ $\quad$ $\cos 60° = 1:2 = 4:8$ $\quad$ $\tan 60° = 4\sqrt{3}:4$

What is the length of *PR*? (1 mark, ★★★)

Label the sides opposite, adjacent and hypotenuse to help you decide whether to use sin, cos or tan.

....

3. Triangle *XYZ* has a right angle at *Y*.

   Angle *YXZ* = 30° $\quad$ Side *XZ* = 20 cm

   What is the length of side *YZ*? (2 marks, ★★★★)

   Draw a diagram of the triangle and put all the information on it.

   ....

4. In a right-angled triangle the ratio between one side and the hypotenuse is $1:\sqrt{2}$

   What are the sizes of the other angles in the triangle? (2 marks, ★★★★)

   ....

Transformations

① On the grid, translate the parallelogram by vector $\begin{pmatrix}-3\\-2\end{pmatrix}$. (2 marks, ★★)

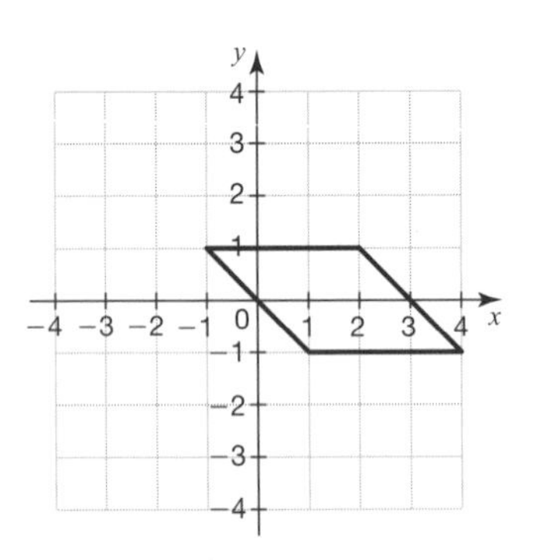

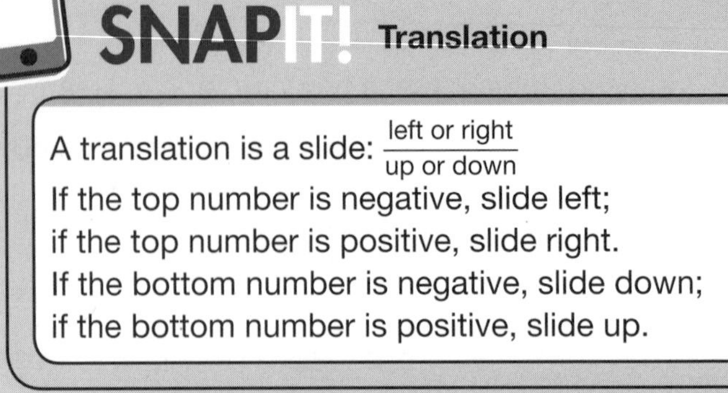

② **a** On the grid below right, rotate the triangle 180° clockwise about (−2, 0).

Label the new triangle A. (2 marks, ★★★)

Trace the triangle. Put your pencil on the point (−2, 0) and rotate the tracing paper 180° clockwise.

b On the grid, enlarge your triangle A by scale factor 3 about centre (−2, 1). (2 marks, ★★★)

Label the new triangle B.

Each point on triangle B must be three times as far from point (-2, 1) as the same point on triangle A.

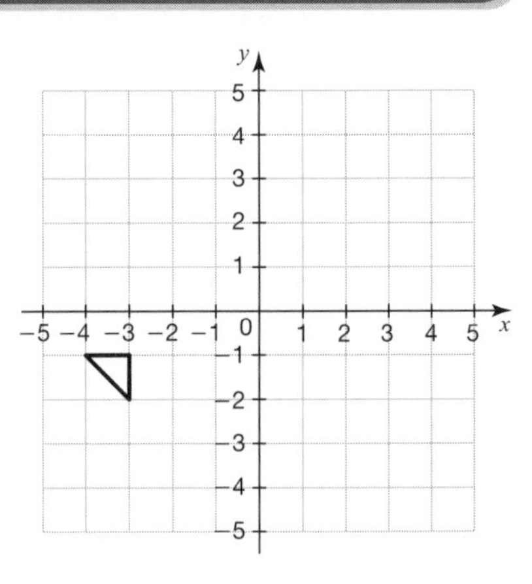

[Total: 4 marks]

③ Describe fully the single transformation that maps trapezium A onto trapezium B. (2 marks, ★★★★)

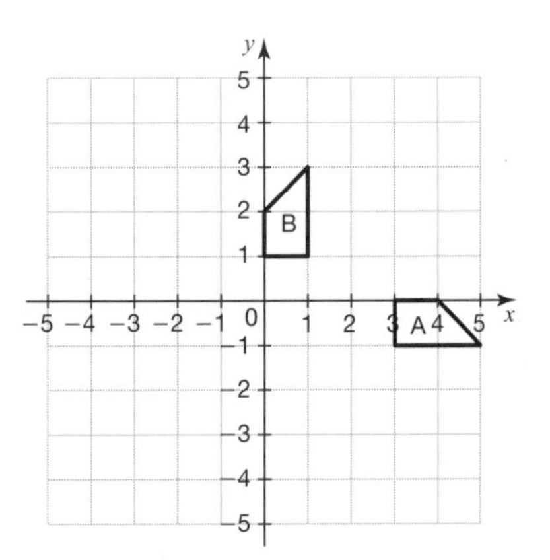

....

....

④ **a** On the grid, rotate shape P 180° about (1, 1). Label the new shape Q. (2 marks, ★★★★)

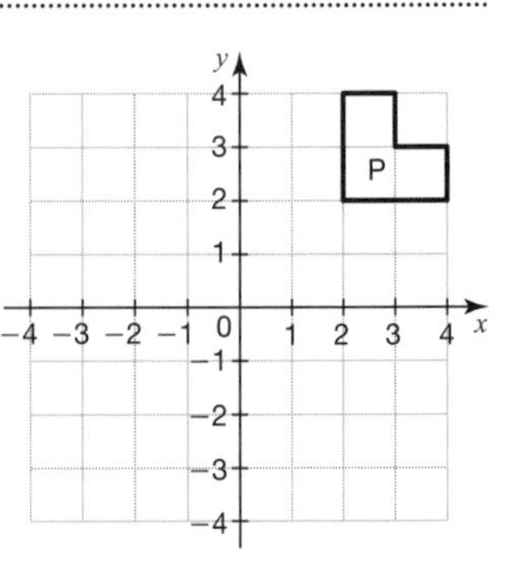

b Reflect shape Q in the line $x = 1$. Label the new shape R. (2 marks, ★★★★)

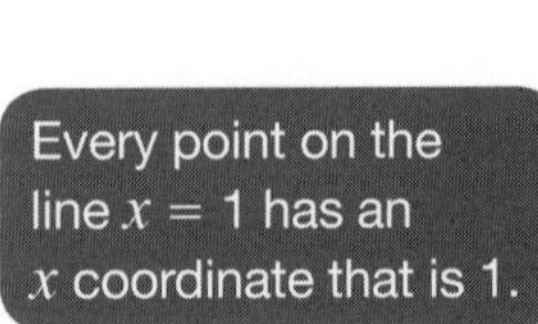

c Describe fully the single transformation that maps shape R onto shape P. (2 marks, ★★★★)

....

[Total: 6 marks]

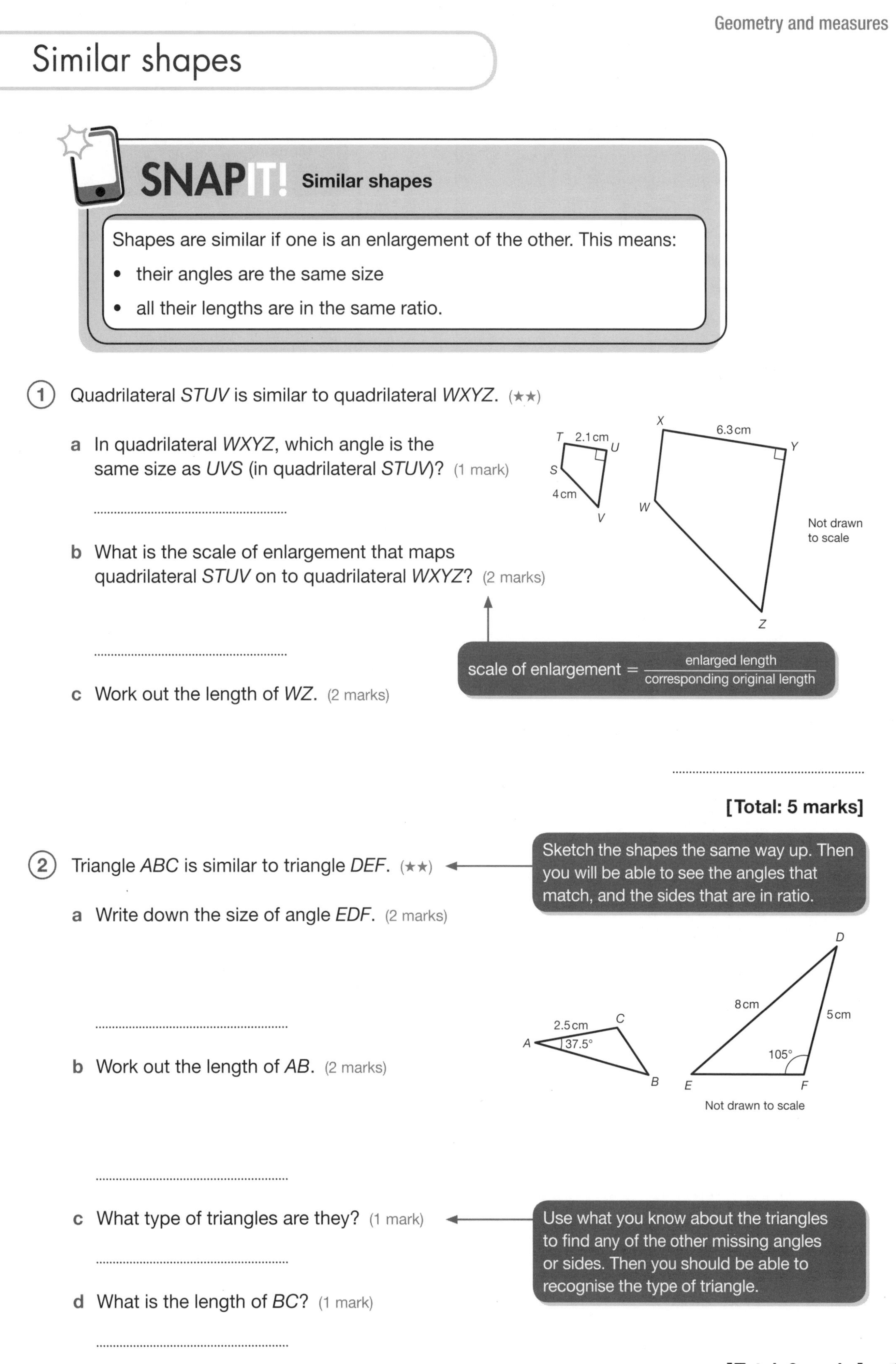

Similar shapes

SNAPIT! Similar shapes

Shapes are similar if one is an enlargement of the other. This means:

- their angles are the same size
- all their lengths are in the same ratio.

(1) Quadrilateral *STUV* is similar to quadrilateral *WXYZ*. (★★)

a In quadrilateral *WXYZ*, which angle is the same size as *UVS* (in quadrilateral *STUV*)? (1 mark)

....

b What is the scale of enlargement that maps quadrilateral *STUV* on to quadrilateral *WXYZ*? (2 marks)

....

scale of enlargement = $\frac{\text{enlarged length}}{\text{corresponding original length}}$

c Work out the length of *WZ*. (2 marks)

....

[Total: 5 marks]

(2) Triangle *ABC* is similar to triangle *DEF*. (★★)

Sketch the shapes the same way up. Then you will be able to see the angles that match, and the sides that are in ratio.

a Write down the size of angle *EDF*. (2 marks)

....

Not drawn to scale

b Work out the length of *AB*. (2 marks)

....

c What type of triangles are they? (1 mark)

Use what you know about the triangles to find any of the other missing angles or sides. Then you should be able to recognise the type of triangle.

....

d What is the length of *BC*? (1 mark)

....

[Total: 6 marks]

③ Triangle *PQR* is similar to triangle *SRT*. (★★★★)

Sketch the two separate triangles. Make sure you sketch them the same way up.

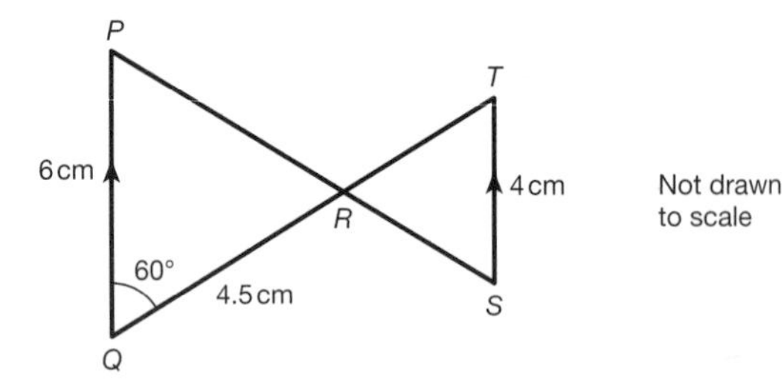

Not drawn to scale

a What is the scale of enlargement that maps triangle *PQR* on to triangle *SRT*? (2 marks)

....

b Work out the length of *RT*. (2 marks)

....

[Total: 4 marks]

④ The lines *AE* and *BD* are parallel. Triangle *ACE* is similar to triangle *BCD*. (★★★★)

a What is the scale of enlargement that maps triangle *ACE* on to triangle *BCD*? (2 marks)

....

b Work out the length *CE*. (1 mark)

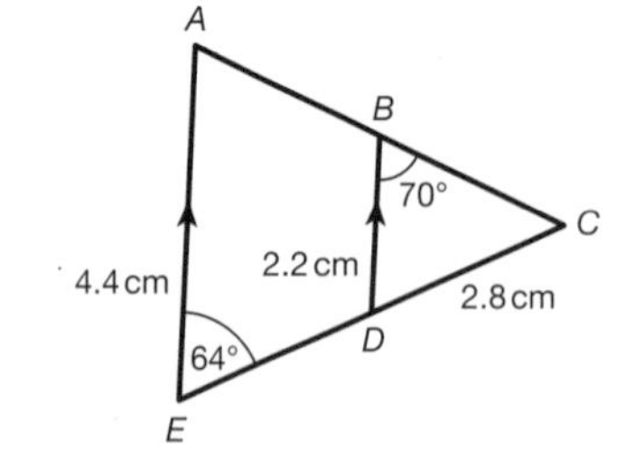

....

c Work out angle *ACE*. (2 marks)

....

[Total: 5 marks]

Vectors

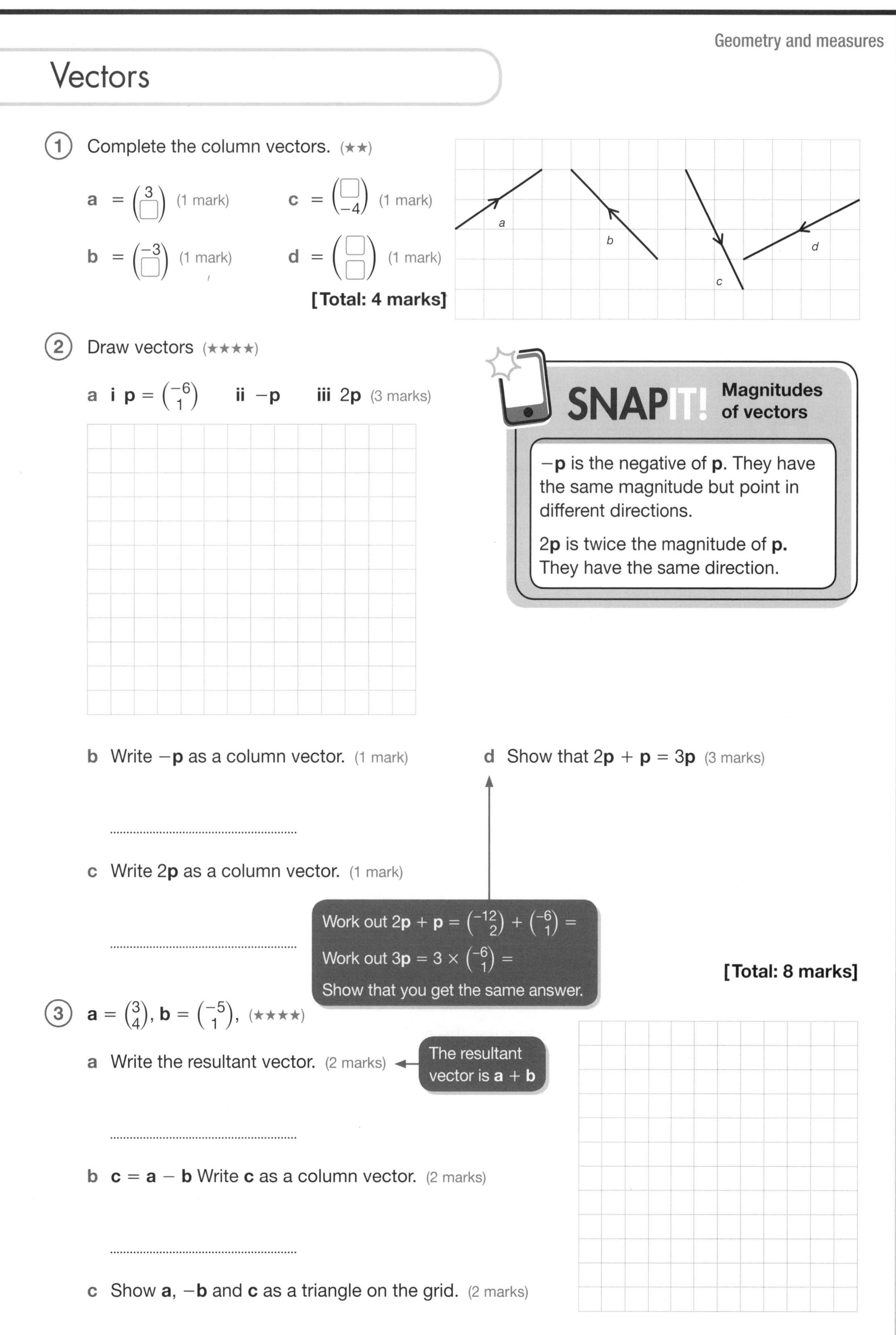

① Complete the column vectors. (★★)

a $= \begin{pmatrix} 3 \\ \square \end{pmatrix}$ (1 mark)

c $= \begin{pmatrix} \square \\ -4 \end{pmatrix}$ (1 mark)

b $= \begin{pmatrix} -3 \\ \square \end{pmatrix}$ (1 mark)

d $= \begin{pmatrix} \square \\ \square \end{pmatrix}$ (1 mark)

[Total: 4 marks]

② Draw vectors (★★★★)

a **i** $\mathbf{p} = \begin{pmatrix} -6 \\ 1 \end{pmatrix}$ **ii** $-\mathbf{p}$ **iii** $2\mathbf{p}$ (3 marks)

SNAP IT! Magnitudes of vectors

$-\mathbf{p}$ is the negative of **p**. They have the same magnitude but point in different directions.

2**p** is twice the magnitude of **p.** They have the same direction.

b Write $-\mathbf{p}$ as a column vector. (1 mark)

....

c Write 2**p** as a column vector. (1 mark)

....

d Show that $2\mathbf{p} + \mathbf{p} = 3\mathbf{p}$ (3 marks)

Work out $2\mathbf{p} + \mathbf{p} = \begin{pmatrix} -12 \\ 2 \end{pmatrix} + \begin{pmatrix} -6 \\ 1 \end{pmatrix} =$

Work out $3\mathbf{p} = 3 \times \begin{pmatrix} -6 \\ 1 \end{pmatrix} =$

Show that you get the same answer.

[Total: 8 marks]

③ $\mathbf{a} = \begin{pmatrix} 3 \\ 4 \end{pmatrix}$, $\mathbf{b} = \begin{pmatrix} -5 \\ 1 \end{pmatrix}$, (★★★★)

a Write the resultant vector. (2 marks)

The resultant vector is **a** + **b**

....

b $\mathbf{c} = \mathbf{a} - \mathbf{b}$ Write **c** as a column vector. (2 marks)

....

c Show **a**, $-\mathbf{b}$ and **c** as a triangle on the grid. (2 marks)

[Total: 6 marks]

Probability
Basic probability

1. There are ten cookery books on a shelf.

   Five are for Indian food.

   Four are for Italian food.

   One is for Chinese food.

   A cookery book is taken at random from the shelf.
   What is the probability it is for Chinese food? (1 mark, ★)

   ....

> successful outcome = cookery book is for Chinese food
>
> probability = $\frac{\text{number of successful outcomes}}{\text{total number of possible outcomes}}$

SNAPIT! Probability questions

A probability question may ask for an answer as a fraction, decimal or percentage.

If it doesn't state 'Give your answer as…', and the question includes fractions, then answer with a fraction; if the question includes decimals, then answer with a decimal, and so on.

If it doesn't state 'Give your answer as…' and there is no fraction, decimal or percentage in the question, then you can choose how to write your answer.

2. The probability it will rain today is 0.6.

   What is the probability it will not rain today? (2 marks, ★★)

   ....

NAILIT!

If the probability of an event occurring is P then the probability of it **not** occurring is $1 - P$.

3. Oliver spins this fair spinner. (★★★)

   a On the probability scale, mark A to show the probability that the spinner will land on a number from 1 to 8. (1 mark)

   b On the probability scale, mark B to show the probability that the spinner will land on a multiple of 3. (2 marks)

   c On the probability scale, mark C to show the probability that the spinner will land on a number greater than 3. (2 marks)

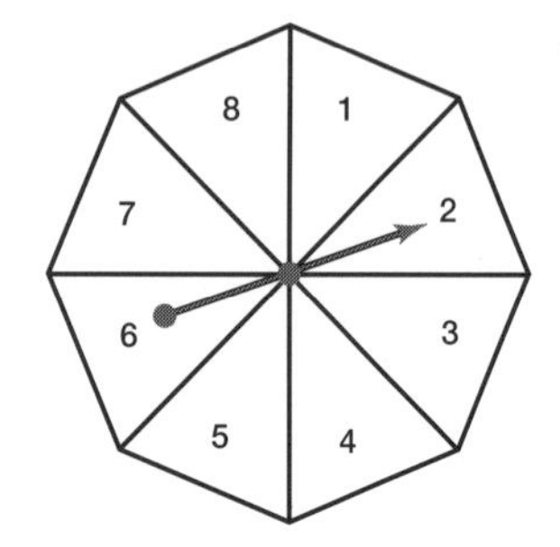

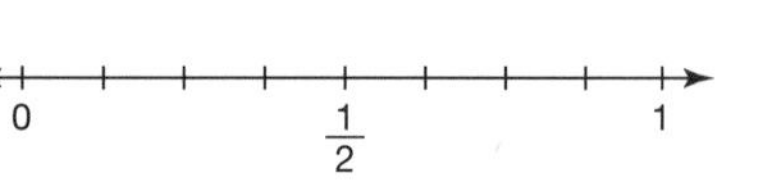

[Total: 5 marks]

When events cannot happen at the same time, then they are mutually exclusive. Their probabilities add to 1.

4 The table shows the probabilities of randomly picking different coloured tickets in a lucky dip.

Outcome	Red	Blue	Green
Probability	$2p - 0.1$	$2p + 0.1$	p

Which colour is most likely? You must show working to justify your answer. (3 marks, ★★★★★)

Write an equation and solve it to find p.

....

Write and solve an equation where all probabilities add to 1.

Two-way tables and sample space diagrams

1. A factory makes wooden bed frames.

   They make singles, doubles and king size.

   They use oak, pine or walnut.

   The two-way table shows the number of beds the factory makes in one week.

   Complete the table. (4 marks, ★★)

	Single	Double	King	Totals
Oak	2			30
Pine		14	17	54
Walnut	1	12		
Totals			32	100

Look for a row or column with only one number missing. Work out that missing number.

2. A coin is flipped and this fair spinner is spun. (★★★)

   a The sample space shows some of the possible outcomes.

A sample space is a type of two-way table. It helps identify all possible outcomes for two independent events.
Events are independent if the outcome of one event does not affect the outcome of another event.

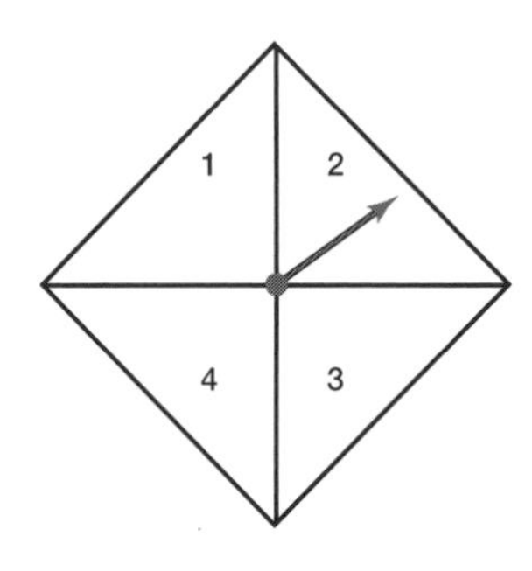

Complete the sample space. (3 marks)

		Spinner			
		1	2	3	4
Coin	**Heads**	1, H			
	Tails			3, T	

b What is the probability of flipping a tail on the coin and spinning a 1 on the spinner? (1 mark)

....

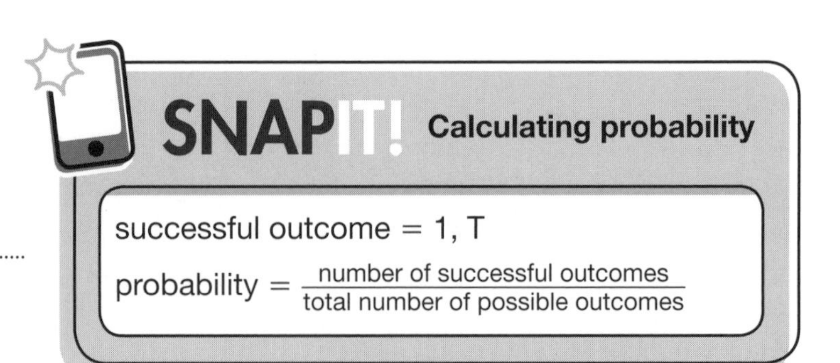

c What is the probability of flipping a head on the coin and a number greater than 1 on the spinner? (1 mark)

....

[Total: 5 marks]

③ There are 120 students in the sixth form.

$\frac{3}{8}$ of the sixth form are boys.

$\frac{1}{5}$ of the sixth form are boys that study sciences.

$\frac{1}{3}$ of the sixth form are girls that do not study sciences. (★★★★)

a Complete the two-way table. (4 marks)

	Study sciences	Do not study sciences	Totals
Boys			
Girls			
Totals			120

b A student is chosen at random. What is the probability the student is a boy who does not study science? (2 marks)

....

NAILIT!

When using a fraction to answer a probability question, always cancel the fraction if you can and write it in its simplest terms.

c A girl is chosen at random. What is the probability she studies science? (2 marks)

....

[Total: 8 marks]

STRETCHIT!

When the number of balls in the lottery was increased from 49 to 59, the chances of winning went down to 1 in 45 000 000, or 0.000002%.

The chances of winning are now even lower than the likelihood of being hit by a meteorite!

How could you use probability to help you make sensible decisions?

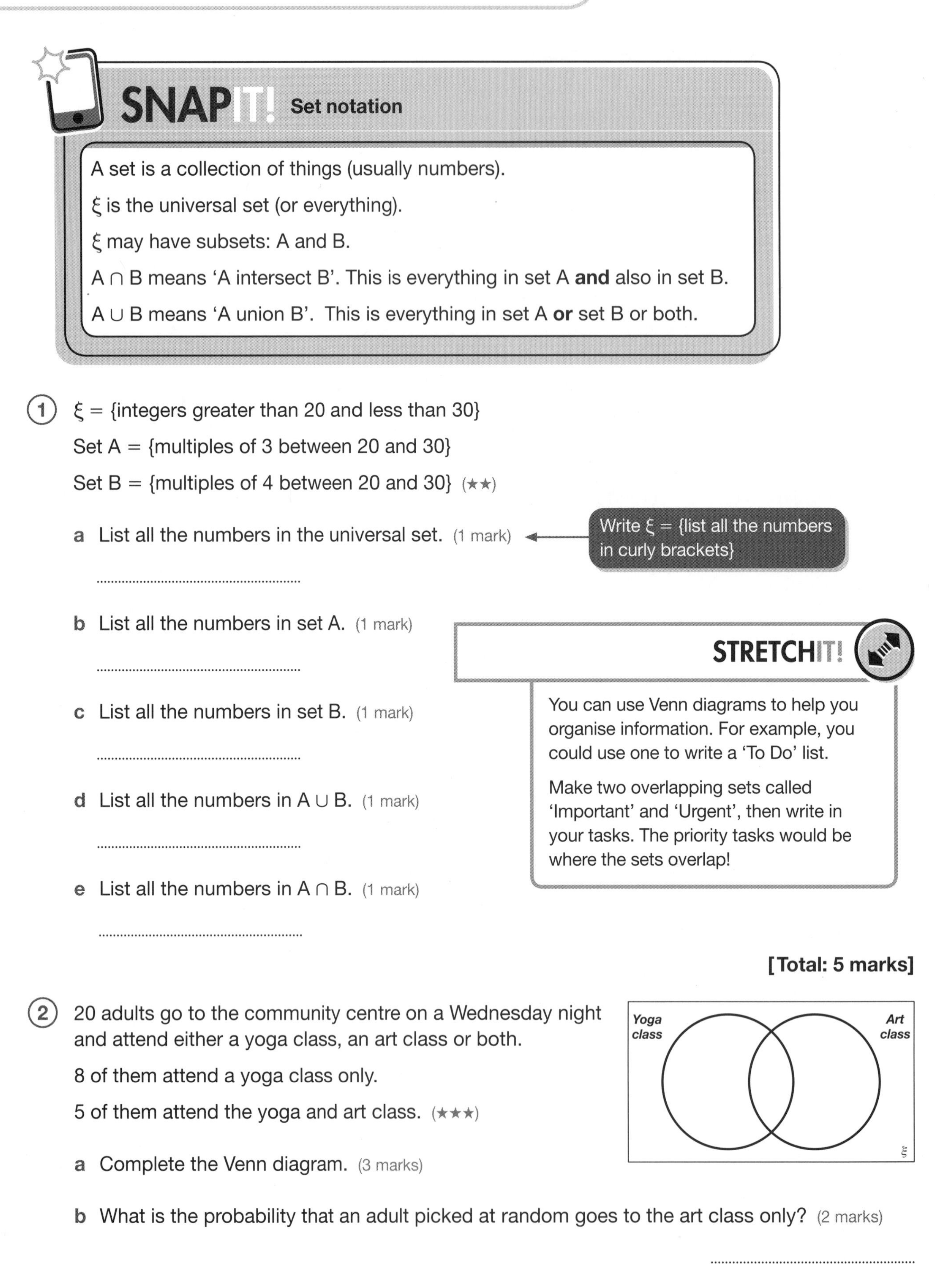

Sets and Venn diagrams

SNAPIT! Set notation

A set is a collection of things (usually numbers).

ξ is the universal set (or everything).

ξ may have subsets: A and B.

$A \cap B$ means 'A intersect B'. This is everything in set A **and** also in set B.

$A \cup B$ means 'A union B'. This is everything in set A **or** set B or both.

1 ξ = {integers greater than 20 and less than 30}

Set A = {multiples of 3 between 20 and 30}

Set B = {multiples of 4 between 20 and 30} (★★)

a List all the numbers in the universal set. (1 mark)

Write ξ = {list all the numbers in curly brackets}

....

b List all the numbers in set A. (1 mark)

....

c List all the numbers in set B. (1 mark)

....

d List all the numbers in $A \cup B$. (1 mark)

....

e List all the numbers in $A \cap B$. (1 mark)

....

STRETCHIT!

You can use Venn diagrams to help you organise information. For example, you could use one to write a 'To Do' list.

Make two overlapping sets called 'Important' and 'Urgent', then write in your tasks. The priority tasks would be where the sets overlap!

[Total: 5 marks]

2 20 adults go to the community centre on a Wednesday night and attend either a yoga class, an art class or both.

8 of them attend a yoga class only.

5 of them attend the yoga and art class. (★★★)

a Complete the Venn diagram. (3 marks)

b What is the probability that an adult picked at random goes to the art class only? (2 marks)

....

c What is the probability that an adult picked at random goes to only one class? (2 marks)

....

[Total: 7 marks]

3. People were surveyed and asked where they bought their food.

S – supermarket

L – local corner shop

F – farmers' market

The Venn diagram shows the responses. (★★★★★)

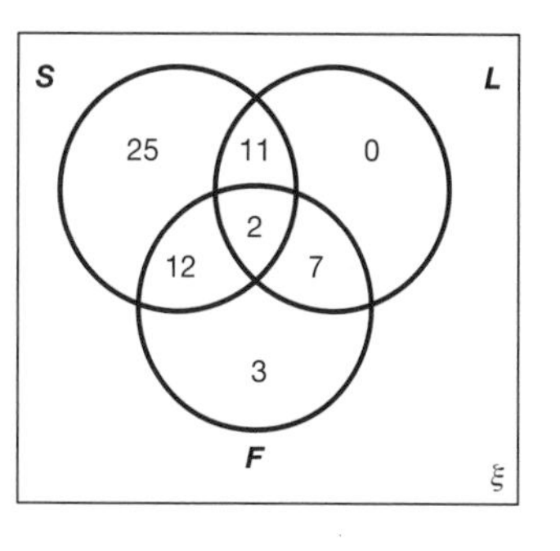

a One of the people surveyed is picked at random.

Work out the probability they bought their food only from a supermarket. (2 marks)

First of all, work out how many people were surveyed: ξ = ☐

....

b One of the people surveyed is picked at random.

Work out P(F). (2 marks)

Don't forget to include everyone in set F.

....

c One of the people surveyed is picked at random. Work out P $(L \cap F)$. (2 marks)

....

d One of the people surveyed is picked at random.

Work out P(S′). (2 marks)

S′ means everyone **not** in set S (i.e. that do not shop at the supermarket at all).

....

[Total: 8 marks]

Frequency trees and tree diagrams

1. There is one black and three white balls in a bag.

   One ball is taken from the bag, its colour is recorded, and then it is put back in the bag.

   A second ball is taken from the bag, its colour is recorded. (★★)

   The events are independent, because the first ball is put back in the bag. This means that the outcome of the first ball being taken from the bag does not affect the outcome of the second ball being taken from the bag.

   a Complete the tree diagram. (3 marks)

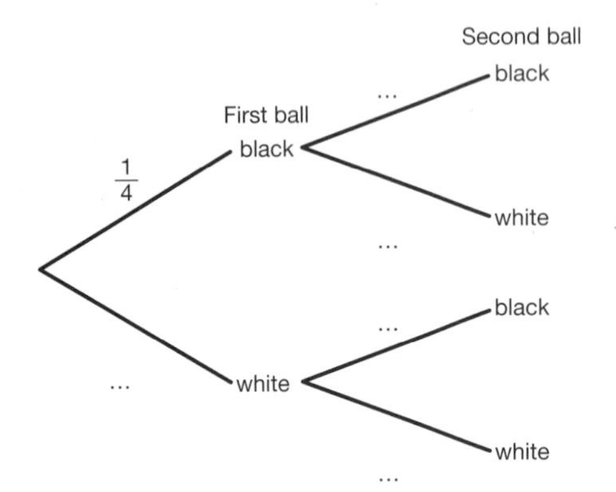

Write the probability for each event on each branch.

The probability of the first ball being black has already been written for you.

   b Work out the probability of the first ball being white and the second ball being black. (2 marks)

   P(W, B) = ...

   c Work out the probability of the first ball being black and the second ball being black. (2 marks)

   P(B, B) = ...

NAIL IT!

To use a tree diagram, read along the branches to find each probability and then multiply them together.

[Total: 7 marks]

2. There are four apples and two oranges in a fruit bowl. Adam takes a piece of fruit from the bowl and eats it. He then takes another piece of fruit for later. (★★★★★)

   a Complete the tree diagram. (3 marks)

   The events are **not** independent, because Adam eats the first piece of fruit. This means that the outcome of his first choice affects the possibilities for his second choice.

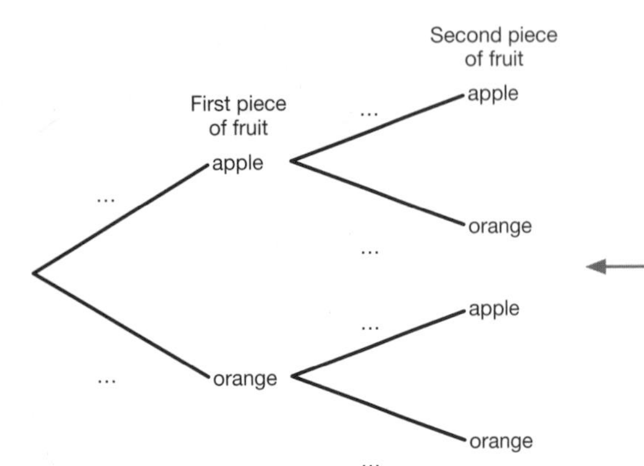

Write the probability for each event on each branch.

Be careful when writing the probabilities for the second piece of fruit taken. There are only six pieces of fruit left in the fruit bowl.

b Work out the probability of Adam eating an apple (call this A) and then taking an orange (O) for later. (2 marks)

P(A, O) = ...

c Work out the probability of Adam eating an orange and then taking an orange for later. (2 marks)

P(O, O) = ...

d What is the probability of the fruit Adam eats being the same as the one he takes for later? (3 marks)

Add together the probabilities for different possibilities:
P (O, O) + P (A, A)

....

[Total: 10 marks]

3 A weather forecaster gives the probability of rain on Saturday as 30% and rain on Sunday as 50%. (★★★★★)

a Draw a tree diagram for the two days, showing the outcomes, 'rain' and 'no rain'. (3 marks)

Write the percentage probabilities as fractions.

b What is the probability of rain on both days? (2 marks)

....

c What is the probability of one day having rain and one day having no rain? Give your answer as a percentage. (2 marks)

....

[Total: 7 marks]

Expected outcomes and experimental probability

① The table shows the results of an experiment with this spinner. (★★★)

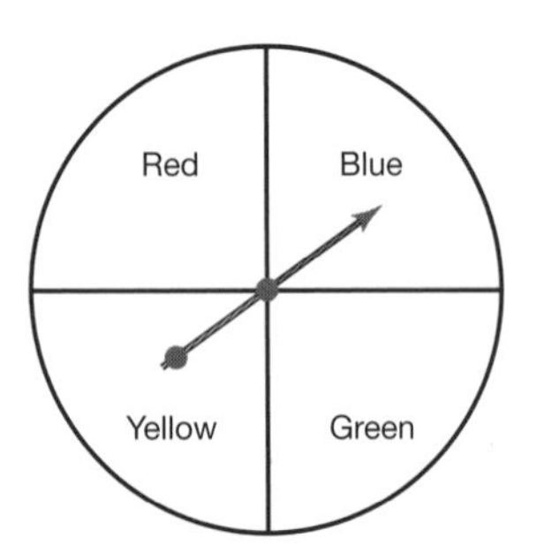

NAILIT!

When writing answers that are fractions, always cancel and give your answer in its simplest form.

Colour	Frequency
Red	12
Blue	13
Green	10
Yellow	15

a How many times in total was the spinner spun? (1 mark)

....

b What is the estimated probability of spinning blue? (2 marks)

....

c What is the estimated probability of spinning yellow? (2 marks)

....

d How many times would you expect to spin green in 100 spins? (2 marks) ← P(Green) × 100

....

[Total: 7 marks]

② The probability of passing a dance exam is 0.75.

A dance school enters 20 students for the exam. How many should they expect to pass? (1 mark, ★★★)

....

(3) The table shows the reasons customers visit a post office in one hour. (★★★★★)

Reason	Frequency
Send a parcel	26
Buy stamps	20
Buy foreign currency	6
Banking services	5
Other	3

a Use the data to estimate the probability that in the next hour someone will buy stamps. (2 marks)

....

The post office has, on average, 450 customers each day.

b Estimate the number of people who buy stamps from the post office each day. (1 mark)

....

c Estimate the number of people who buy foreign currency from the post office each day. (3 marks)

....

d Estimate the number of people who use the post office for banking services each week. (The post office is open six days per week). (3 marks)

....

[Total: 9 marks]

Statistics
Data and sampling

1. A large bakery produces 650 bread rolls each day. The head baker wants to check the quality of a sample of the rolls. How many should she check? Explain. (2 marks, ★)

A good sample is 10% of the whole population. The population is the whole group you are interested in. In this case it is all 650 bread rolls.

....

....

2. A town councillor wishes to investigate how many people use the buses. The population of the town is 25 000. He takes a sample of 50 people in the town centre. Give two reasons why this is not a good sample. (2 marks, ★★★)

....

....

....

NAILIT!

When asked to 'Explain', write your answer, then write 'because'. Finish the sentence with a reason for your answer.

3. 400 people have accepted an invitation to attend a big street party.

Sam is organising cakes.

She emails a sample of 45 people who are coming to the party.

She asks each one to tell her what type of cake they like.

The table shows information about her results. (★★★★★)

Cakes	Number of attendees
Lemon cake	15
Fruit cake	4
Carrot cake	9
Chocolate cake	17

STRETCHIT!

What would Sam do if someone said they didn't eat cake? Consider how she might change the survey to allow for that.

a Work out how many carrot cakes Sam should make. (2 marks)

....

b Write down any assumptions you make and explain how this could affect your answer. (2 marks)

....

....

....

....

....

[Total: 4 marks]

Frequency tables

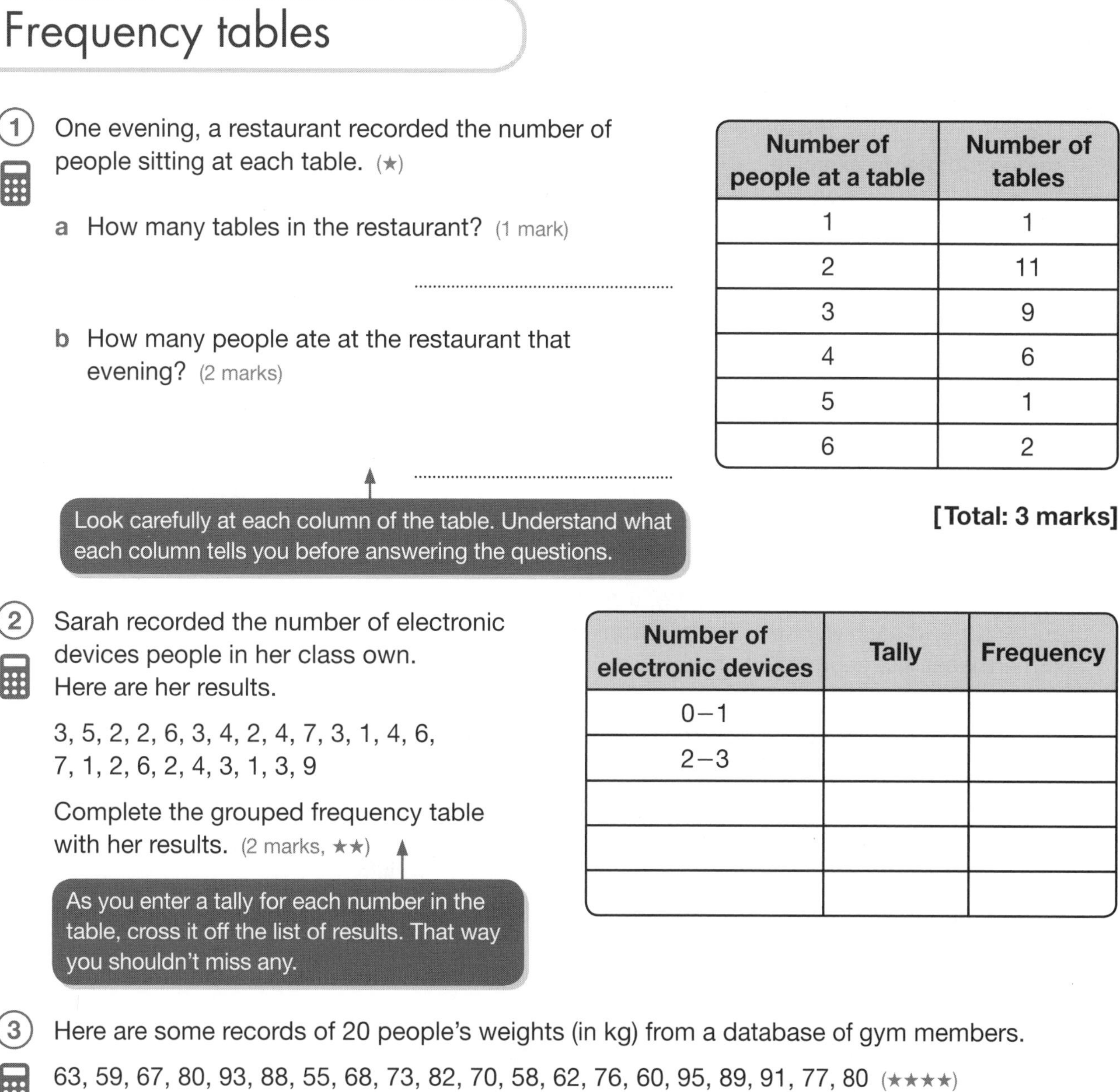

1 One evening, a restaurant recorded the number of people sitting at each table. (★)

Number of people at a table	Number of tables
1	1
2	11
3	9
4	6
5	1
6	2

a How many tables in the restaurant? (1 mark)

....

b How many people ate at the restaurant that evening? (2 marks)

....

Look carefully at each column of the table. Understand what each column tells you before answering the questions.

[Total: 3 marks]

2 Sarah recorded the number of electronic devices people in her class own. Here are her results.

3, 5, 2, 2, 6, 3, 4, 2, 4, 7, 3, 1, 4, 6, 7, 1, 2, 6, 2, 4, 3, 1, 3, 9

Complete the grouped frequency table with her results. (2 marks, ★★)

As you enter a tally for each number in the table, cross it off the list of results. That way you shouldn't miss any.

Number of electronic devices	Tally	Frequency
0–1		
2–3		

3 Here are some records of 20 people's weights (in kg) from a database of gym members.

63, 59, 67, 80, 93, 88, 55, 68, 73, 82, 70, 58, 62, 76, 60, 95, 89, 91, 77, 80 (★★★★)

a Is this data discrete or continuous? (1 mark)

....

b Design and complete a grouped frequency table to record the masses. (3 marks)

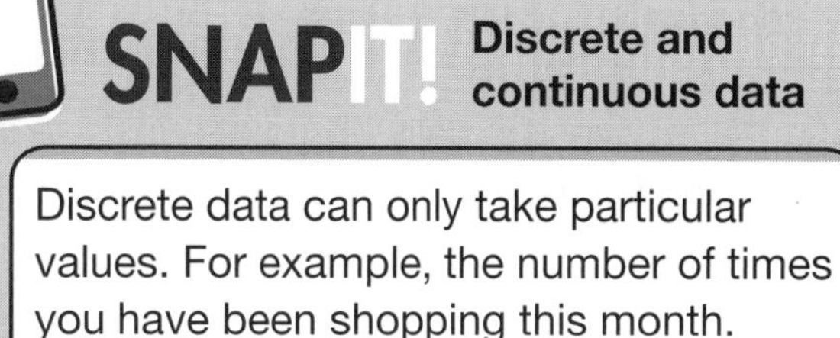

Discrete and continuous data

Discrete data can only take particular values. For example, the number of times you have been shopping this month. It can be organised using groups like 1–5, 6–10, 11–15, etc.

Continuous data is measured and can take any value. For example, a length. Continuous data is organised using ranges like $1 \leq \text{length} < 5$; $6 \leq \text{length} < 10$; $11 \leq \text{length} < 15$, etc.

[Total: 4 marks]

Bar charts and pictograms

1. The bar chart shows the ways that some university students prefer to exercise. (★★)

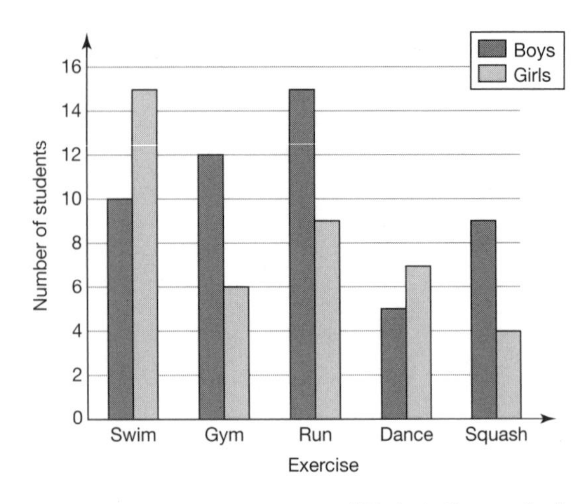

   a How many more boys than girls prefer to play squash? (1 mark)

   ....

   b How many girls were surveyed? (2 marks)

   ....

[Total: 3 marks]

2. The table shows the number of shirts, trousers and suits sold each weekend for four weekends in menswear in a department store.

	Suits	Shirts	Trousers
Weekend 1	10	6	6
Weekend 2	7	15	20
Weekend 3	10	7	10
Weekend 4	7	12	20

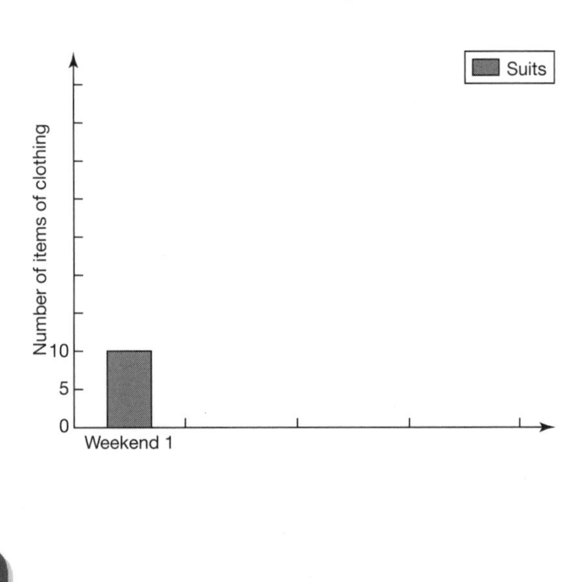

Complete the composite bar chart. (3 marks, ★★★)

The question asks for a composite bar chart. In this type of bar chart, bars for each category (suits, shirts and trousers) are stacked on top of each other.

3. The pictogram shows average daily hours of sunshine in June for three different counties in the UK. (★★★)

Average daily hours of sunshine in June

3 hours of sunshine

Yorkshire

Inverness-shire

Kent

   a What average daily hours of sunshine in June do they get in Kent? (1 mark)

   ....

   b How many more average daily hours of sunshine does Yorkshire get than Inverness-shire? (1 mark)

   ....

For pictograms, always look carefully at the key.

[Total: 2 marks]

Pie charts

1. 300 people were asked what type of movie they last saw.

The pie chart shows their answers. (★)

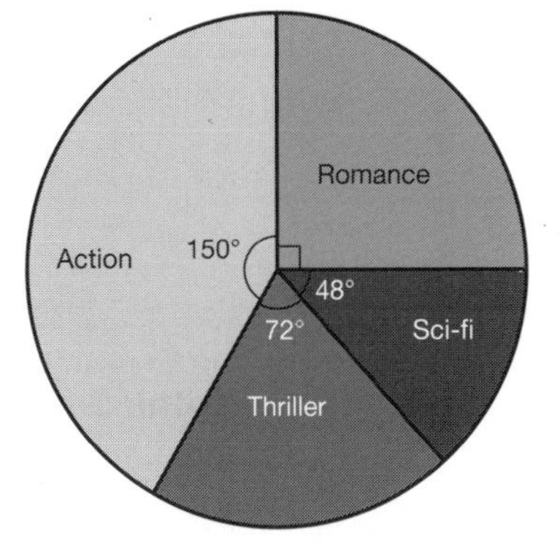

a How many people last saw an action movie? (1 mark)

....

b How many did **not** see a romance movie? (2 marks)

....

[Total: 3 marks]

2. In one day 250 people visited a castle. The table shows information about whether they were children, adults or senior citizens.

Children	Adults	Senior citizens
80	155	15

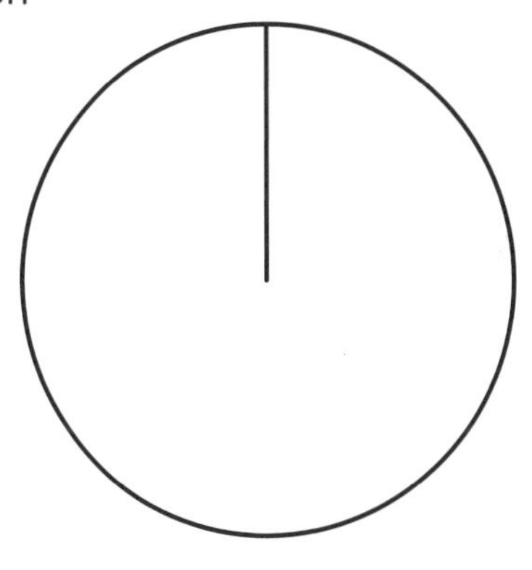

a Draw an accurate pie chart for this information. (3 marks, ★★★)

Work out the fraction that were children. That is the fraction of degrees of the circle you should draw and label children. (Sometimes it may not be a whole number of degrees. Then you will need to round to the nearest degree.)

3. Some 10-year-old children were surveyed about the time they usually go to bed.

Here is a pie chart of the results.

12 of the children answered 8.30 – 9 pm.

How many children were surveyed? (2 marks, ★★★★)

Measure the angle for 8.30 – 9 pm. How many degrees is one child?

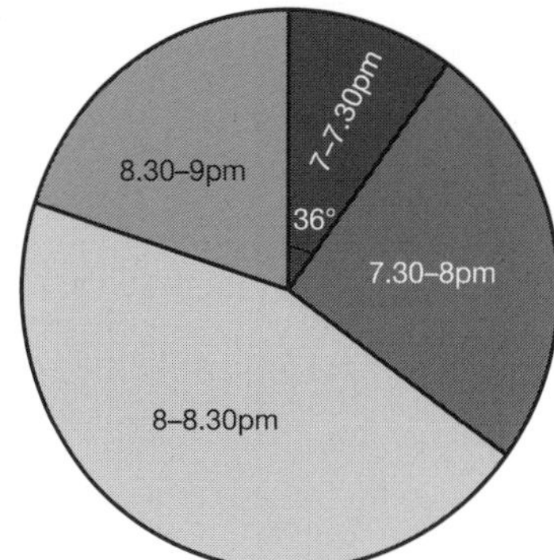

....

Stem and leaf diagrams

1. The stem and leaf diagram shows the height in centimetres of different tomato plants. (★)

2	3 9
3	0 1 1
4	2 5 6 7
5	1 4 8 8 8 9
6	0 3 6 9
7	1

Key
2 | 3 represents 23 cm

a What is the height of the tallest tomato plant? (1 mark)

....

b How many tomato plants were measured? (1 mark)

....

Look carefully at the key. It tells you that the stem (2, 3, 4, 5, 6, 7) is the tens and the leaf is the units.

[Total: 2 marks]

2. Hal recorded the distances (in metres) Year 11 students jumped in long jump on sports day. This back-to-back stem and leaf diagram shows the results. (★★)

Girls		Boys
80 71 09	2	03 94
74 65 13 02	3	42 44 82 99
06	4	11 26

Key:
Girls: 09|2 means 2.09 m
Boys: 2|03 means 2.03 m

a What is the difference between the longest and shortest jump by the boys? (1 mark)

....

b How many girls jumped more than 3.5 m? (1 mark)

....

[Total: 2 marks]

3. The table shows the amounts spent by Year 9s and Year 10s one break-time in the school tuckshop.

Year 9s	80p	£1.30	£2.65	95p	50p	£1.75	£3.05	£2
Year 10s	£1.50	£2.70	75p	£3.10	65p	£1.85		

Draw a stem and leaf diagram for this information. (4 marks, ★★★)

Draw the table again, putting all the amounts for each year in order. As you do this, change the pence (p) to pounds (£).

NAIL IT!

When working with measures, it often helps to work only in one unit. For example, if some data is in £ and some in pence, choose to work only in £ or only in pence; if some data is in cm and some in mm, choose to work only in cm or only in mm.

Measures of central tendency: mode

SNAP IT! Mode

The mode is the most common value or values.

1. Here are the times (to the nearest minute) people waited in a queue to buy cinema tickets.

   2 4 6 3 3 5 4 4 3 6

   What is the modal time? (1 mark, ★)

   ....

2. This table shows the ages of students in a drama club.

   What is the modal class? (1 mark, ★★)

   ....

Age of students, a (years)	Frequency
$10 < a \leq 11$	5
$11 < a \leq 12$	12
$12 < a \leq 13$	17
$13 < a \leq 14$	10
$14 < a \leq 15$	6

3. People were surveyed to find out how much money they were carrying in their wallet.

   These are the results.

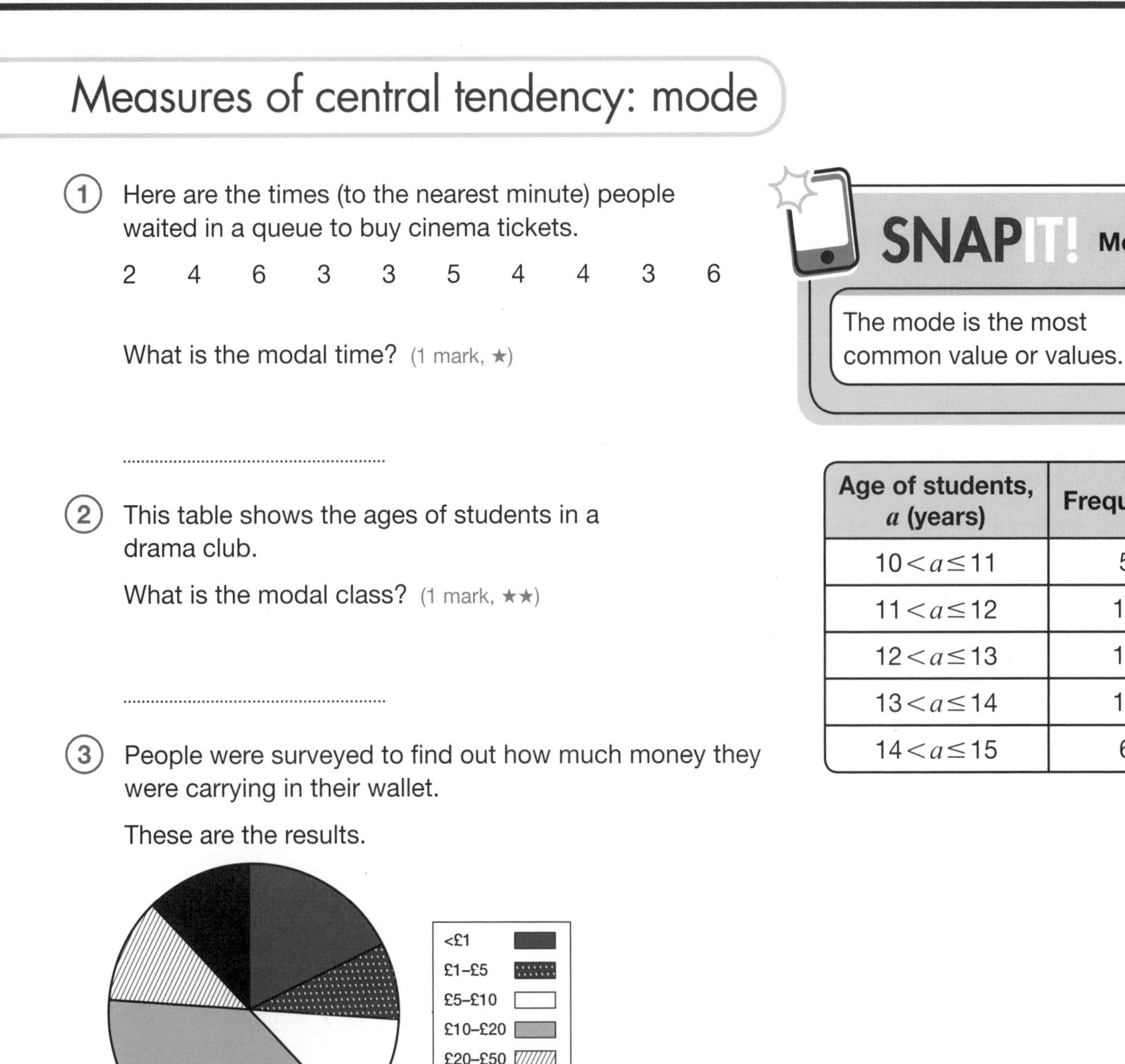

   What is the modal amount? (1 mark, ★★★)

   ....

4. This stem and leaf diagram shows the weight (in kg) of dogs in a kennel.

   What is the modal weight? (1 mark, ★★★)

   ....

Stem	Leaf
0	3 4
1	2 5 8 8
2	1 5 5 5 7
3	0 4 6 9 9
4	1 1 2 6 7 7
5	0 3 5

Key
1 | 2 represents 12 kg

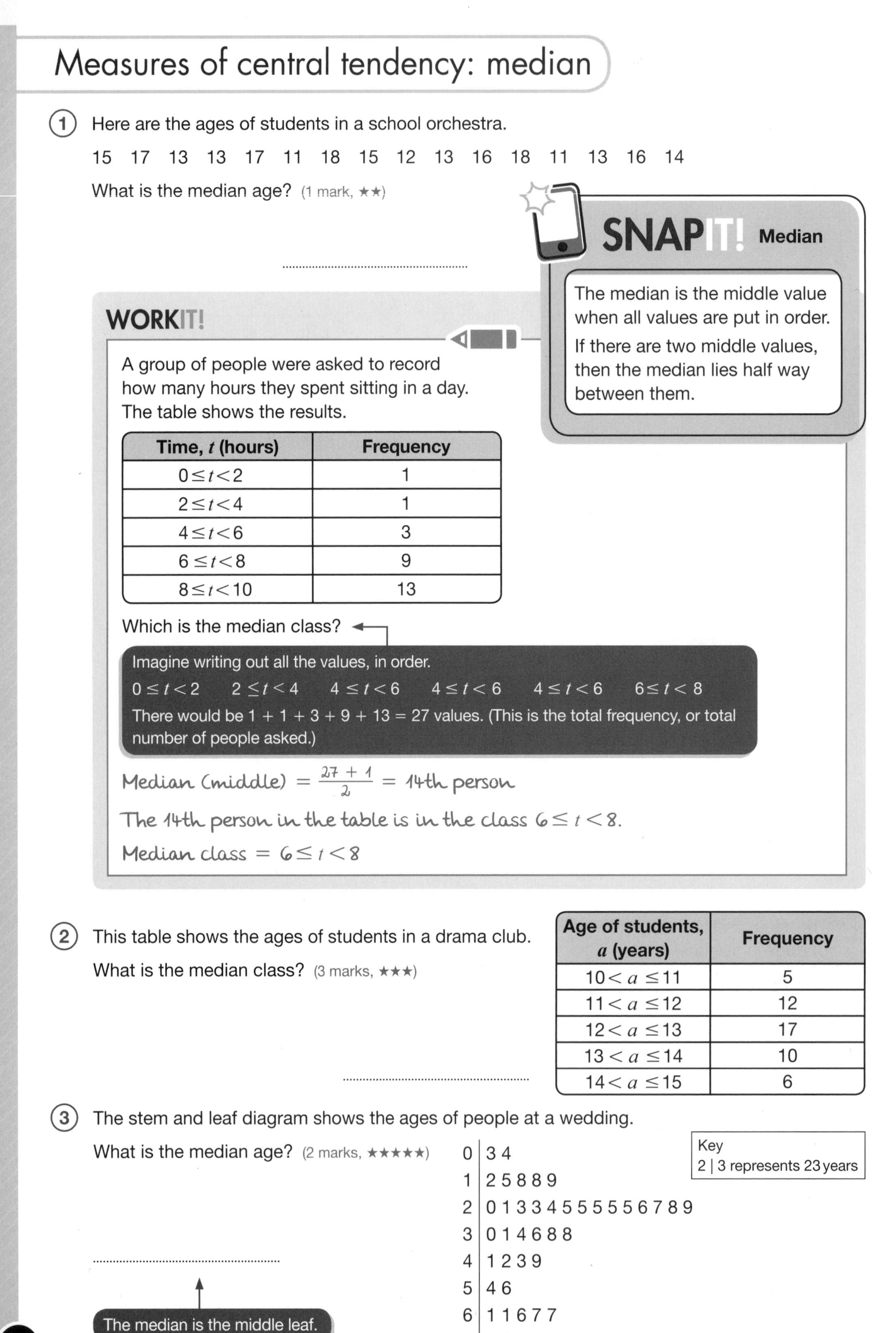

Measures of central tendency: median

1. Here are the ages of students in a school orchestra.

15 17 13 13 17 11 18 15 12 13 16 18 11 13 16 14

What is the median age? (1 mark, ★★)

....

SNAPIT! Median

The median is the middle value when all values are put in order.

If there are two middle values, then the median lies half way between them.

WORKIT!

A group of people were asked to record how many hours they spent sitting in a day. The table shows the results.

Time, t (hours)	Frequency
$0 \le t < 2$	1
$2 \le t < 4$	1
$4 \le t < 6$	3
$6 \le t < 8$	9
$8 \le t < 10$	13

Which is the median class?

Imagine writing out all the values, in order.

$0 \le t < 2$ $2 \le t < 4$ $4 \le t < 6$ $4 \le t < 6$ $4 \le t < 6$ $6 \le t < 8$

There would be 1 + 1 + 3 + 9 + 13 = 27 values. (This is the total frequency, or total number of people asked.)

Median (middle) $= \frac{27 + 1}{2}$ = 14th person

The 14th person in the table is in the class $6 \le t < 8$.

Median class $= 6 \le t < 8$

2. This table shows the ages of students in a drama club.

What is the median class? (3 marks, ★★★)

Age of students, a (years)	Frequency
$10 < a \le 11$	5
$11 < a \le 12$	12
$12 < a \le 13$	17
$13 < a \le 14$	10
$14 < a \le 15$	6

....

3. The stem and leaf diagram shows the ages of people at a wedding.

What is the median age? (2 marks, ★★★★★)

Stem	Leaf
0	3 4
1	2 5 8 8 9
2	0 1 3 3 4 5 5 5 5 5 6 7 8 9
3	0 1 4 6 8 8
4	1 2 3 9
5	4 6
6	1 1 6 7 7
7	0 3

Key
2 | 3 represents 23 years

....

The median is the middle leaf.

Measures of central tendency: mean

1. Here are the ages of children in a family.

   6 7 11 13 18

   What is their mean age? (1 mark, ★★)

   ....

SNAPIT! Mode

$$\text{mean} = \frac{\text{sum of all values}}{\text{number of values}}$$

2. The bar chart shows the number of bedrooms in houses on a newly built estate.

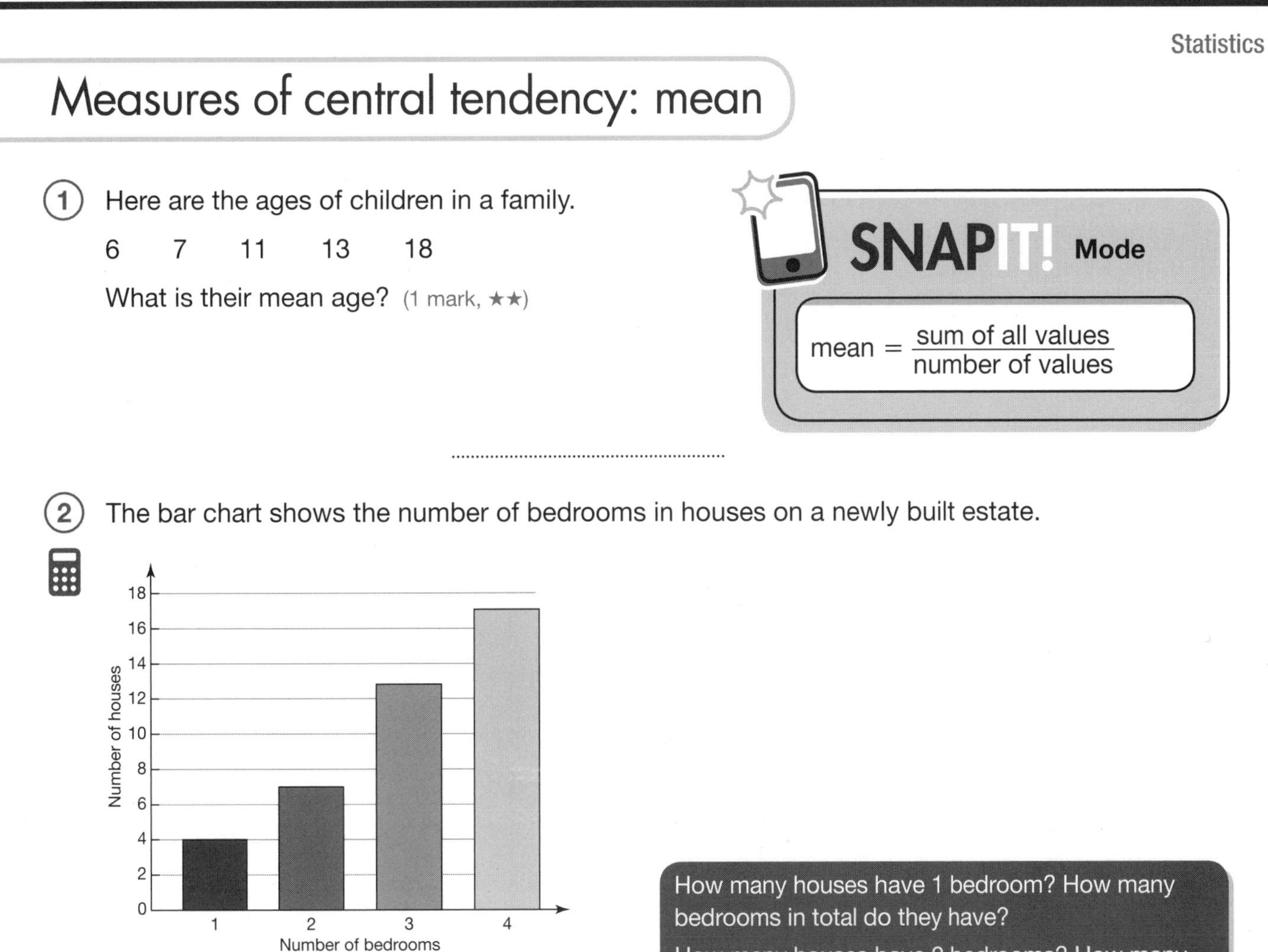

Work out the mean number of bedrooms. (3 marks, ★★★★)

How many houses have 1 bedroom? How many bedrooms in total do they have?

How many houses have 2 bedrooms? How many bedrooms in total do they have?

Repeat these calculations for each number of bedrooms to work out the sum of all the bedrooms.

Divide by the total number of houses.

....

STRETCHIT!

What would you like to find out about? It's a serious question, because the whole point of maths is to work out something you want to know.

Examples in books are written to help you learn the skills. In real life, you write your own questions – and are genuinely interested in the answers!

3 The frequency table shows the number of holidays a company's employees have taken in a year.

Work out the mean number of holidays for all employees. (3 marks, ★★★★)

Number of holidays	Frequency
0	4
1	21
2	9
3	2

NAILIT!

Think about the information you are working with. Does it make sense to give an answer that is a decimal or a fraction? If not, then round to the nearest whole number.

How many people took 0 holidays? How many holidays in total is that?
How many people took 1 holiday? How many holidays in total is that?
Put an extra column on the frequency table to help you work this out:
number of holidays × frequency
Work out the sum of all the holidays.
Divide by the total number of employees (total frequency).

....

4 The grouped frequency table shows the ages of patients waiting to be seen in a hospital accident and emergency department.

What is the mean age of the patients? (2 marks, ★★★)

Create two new columns to help you work out the answer.
Age of patients (a) **Midpoint** **Frequency** **Midpoint × Frequency**

Age of patients (a)	Frequency
$0 < a \leq 10$	3
$10 < a \leq 20$	18
$20 < a \leq 30$	6
$30 < a \leq 40$	11
$40 < a \leq 50$	10
$50 < a \leq 60$	19
$60 < a \leq 70$	16
$70 < a \leq 80$	17

....

Range

1. The bar chart shows the ages of children in a nursery. (★)

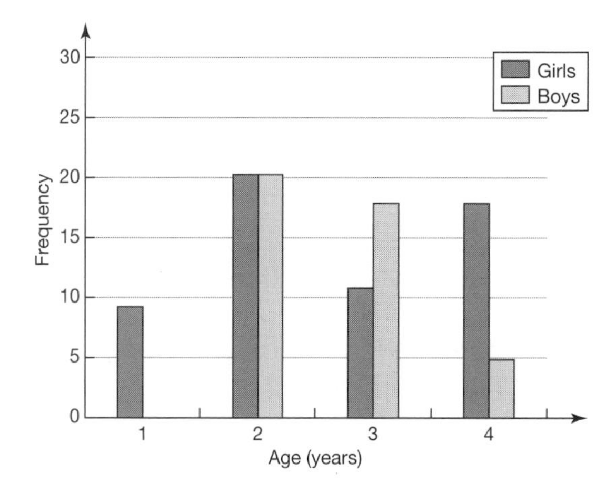

   a What is the range of ages of the boys? (1 mark)

   ....

   b What is the range of ages of the girls? (1 mark)

   ....

SNAPIT! Range

range = highest value − lowest value
Don't forget to include units when giving your answer. For example year, °C, etc.

[Total: 2 marks]

2. The table shows average midday temperatures for ten weeks during the summer. (★★)

Week	1	2	3	4	5	6	7	8	9	10
Temperature °C (Resort A)	17	17	18	21	25	20	21	18	16	16
Temperature °C (Resort B)	13	15	18	23	27	27	28	22	16	14

   a What is the range in temperatures for Resort A? (1 mark)

   ....

   b What is the range in temperatures for Resort B? (1 mark)

   ....

[Total: 2 marks]

3. Here are the profit figures for two small businesses. (★★)

	Year 1	Year 2	Year 3	Year 4
Business A	£23 561	£30 485	£39 210	£45 816
Business B	£32 820	£40 328	£17 894	£63 248

   a What is the range and mean yearly profit for Business A? (2 marks)

   ....

   b What is the range and mean yearly profit for Business B? (2 marks)

   ....

   c Which business is doing better? Explain. (2 marks)

There are 2 marks for part c, so give two reasons for your choice. Use the information you have found in parts a and b.

   ....

   ....

[Total: 6 marks]

Comparing data using measures of central tendency and range

1. The time taken for the same bus journey each morning, Monday to Saturday, is recorded in the table below.

Monday	Tuesday	Wednesday	Thursday	Friday	Saturday
32 minutes	30 minutes	39 minutes	32 minutes	43 minutes	31 minutes

The same journey can be made by train. The time taken for this train journey, Monday to Saturday, is recorded in the table below. (★★)

Monday	Tuesday	Wednesday	Thursday	Friday	Saturday
16 minutes	24 minutes	18 minutes	26 minutes	70 minutes	17 minutes

a Find the mean time for the bus journey. (2 marks)

....

b Find the range for the bus journey. (1 mark)

....

c Find the mean time for the train journey. (2 marks)

....

d Find the range for the train journey. (1 mark)

....

e Is it better to get the bus or the train? Explain. (2 marks)

....

....

[Total: 8 marks]

(2) Two council officers record the ages of people at a playground as

5 4 33 4 8 26 3 7 30 7 8 5 3 39 (★★★★)

a One council officer finds the mean to report the average age of playground users. Explain why this is not the best average to use. (2 marks)

....

....

b The other council officer says that they should use the mode to report the average age of playground users. Explain why this is not the best average to use. (2 marks)

....

....

c The council officers should use the median. Find the median. (1 mark)

....

[Total: 5 marks]

NAIL IT!

When asked to 'explain', use calculations to support your explanation.

(3) There are 20 students in a class. The class teacher records the number of days each student is absent due to sickness in one term:

0 0 0 1 0 1 2 0 5 0 0 1 0 24 2 1 0 0 0 3

Which average should the class teacher use to describe absences due to sickness in her class?

Explain why she should use this average. (3 marks, ★★★★)

....

....

....

NAIL IT!

Look at the number of marks a question is worth. If it is worth 4 marks, then try to include four calculations and/or statements.

Time series graphs

1. The time series graph shows the weight, in grams, of a male kitten for the first seven days of his life. (★★★)

SNAPIT! Time series graph

A time series graph shows how data changes over time. Time is usually plotted on the horizontal (x) axis.

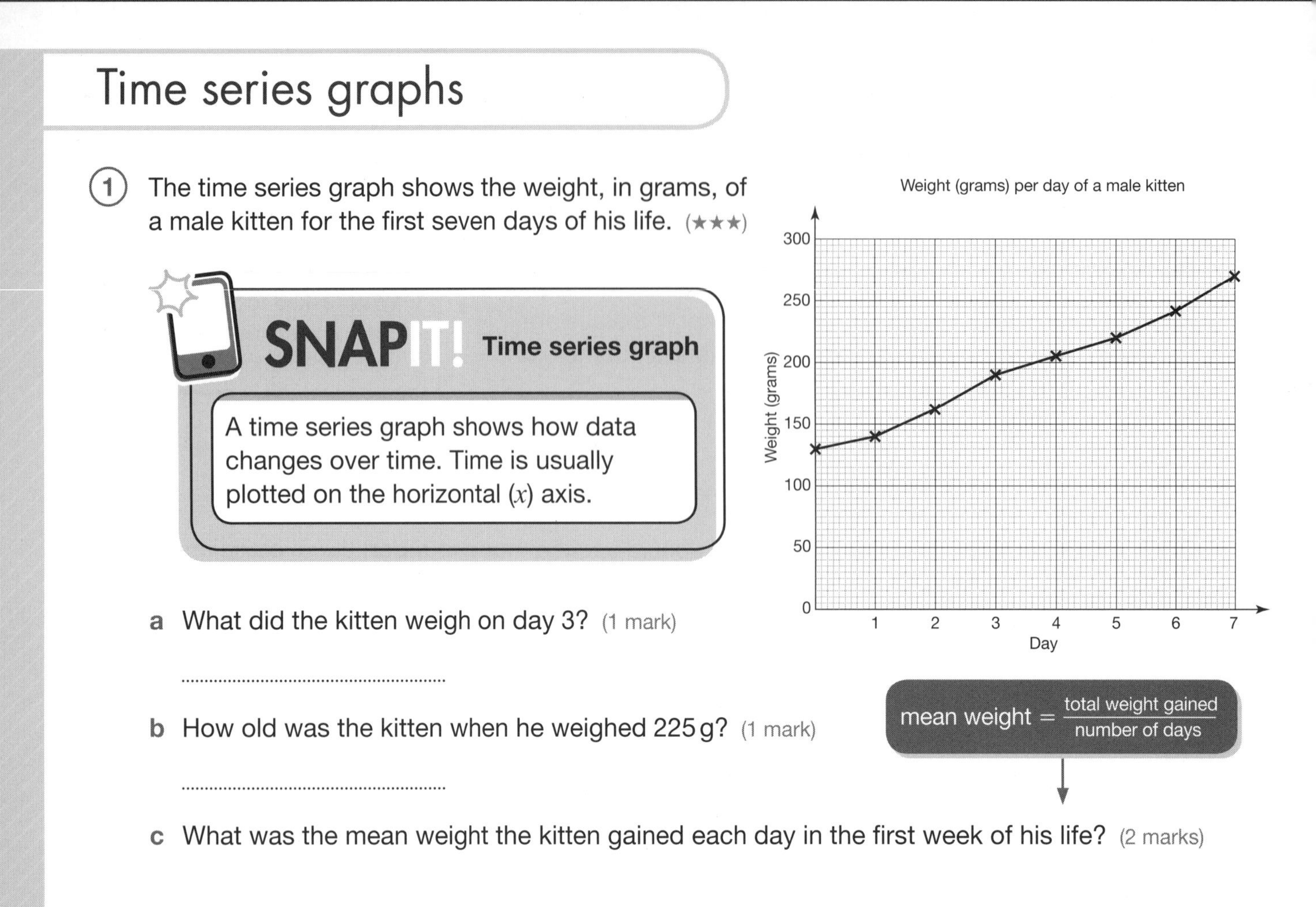

a What did the kitten weigh on day 3? (1 mark)

....

b How old was the kitten when he weighed 225 g? (1 mark)

....

$$\text{mean weight} = \frac{\text{total weight gained}}{\text{number of days}}$$

c What was the mean weight the kitten gained each day in the first week of his life? (2 marks)

....

[Total: 4 marks]

2. A campsite opens for the season in March. The table shows the number of campers on the campsite for the first six months. (★★★)

Month	Mar	Apr	May	Jun	Jul	Aug
Number of campers	30	62	125	80	108	150

a Draw a time series graph to represent the data. (3 marks)

NAILIT!

1. Write a title for your graph.
2. Choose a sensible scale for the number of campers (going up in steps of two will make it a very big graph; going up in steps of 100 will make it a very small graph).
3. Label your axes.
4. Plot your points carefully.
5. Join your points with straight lines.

b What do you notice about May? Give two different sensible reasons for this happening. (3 marks)

....

....

[Total: 6 marks]

(3) Two new branches of a DIY store open in two different towns. The table shows the sales figures for each new DIY store in its first eight months. (★★★★★)

Month	1	2	3	4	5	6	7	8
Branch 1 sales (£)	£6450	£11 740	£16 690	£25 870	£24 920	£25 560	£26 100	£25 930
Branch 2 sales (£)	£4390	£6210	£5950	£20 380	£23 820	£27 460	£32 460	£37 590

a Draw a time series graph for each shop's sales on the same axes. (3 marks)

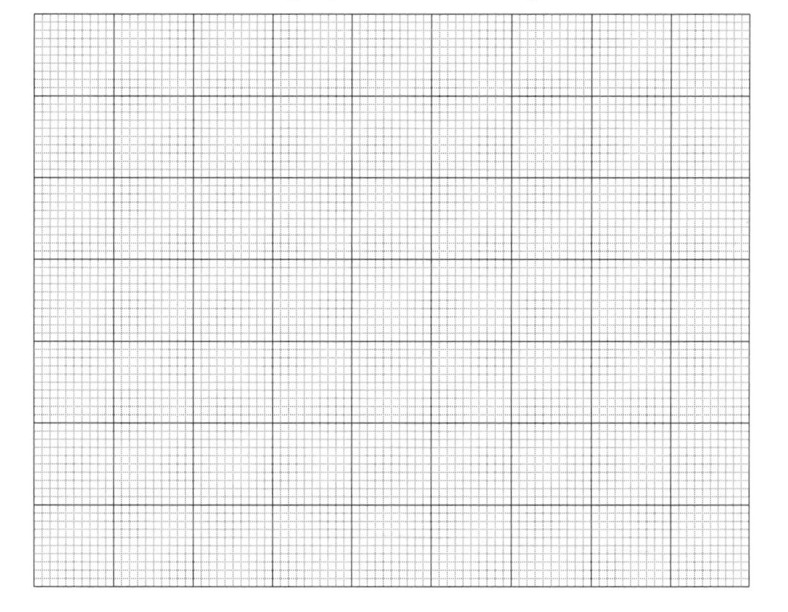

b Describe the trend in sales for each branch for the eight months. (4 marks)

There are 4 marks for part **b**, so you need to write at least two things about Branch 1; and at least two things about Branch 2.

....

....

....

....

[Total: 7 marks]

Scatter graphs

1. Here are two scatter diagrams.

Scatter diagram A shows temperature (°C) and pairs of flip flops sold.

Scatter diagram B shows temperature (°C) and pairs of wellington boots sold. (★)

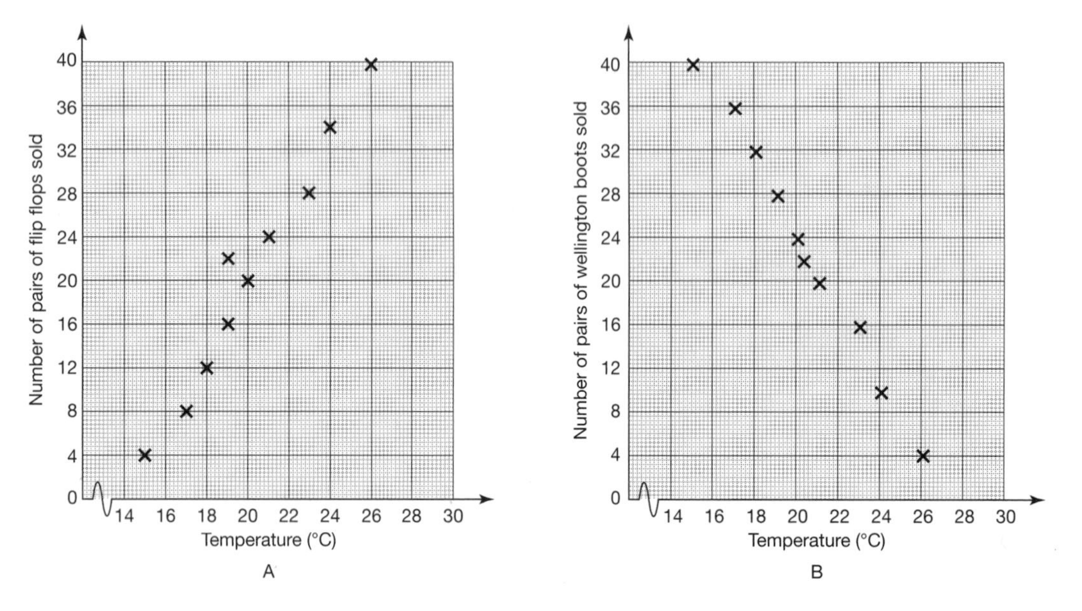

a Describe the correlation for scatter diagram A. Explain. (2 marks)

....

b Describe the correlation for scatter diagram B. Explain. (2 marks)

....

[Total: 4 marks]

2. The table shows the percentage achieved by the same students in a maths test and a physics test. (★★★)

Maths test (%)	45	37	55	70	0	62	48	59	89	30	55	43
Physics test (%)	40	43	64	79	58	66	52	65	80	42	60	40

a Draw a scatter diagram for this data. (3 marks)

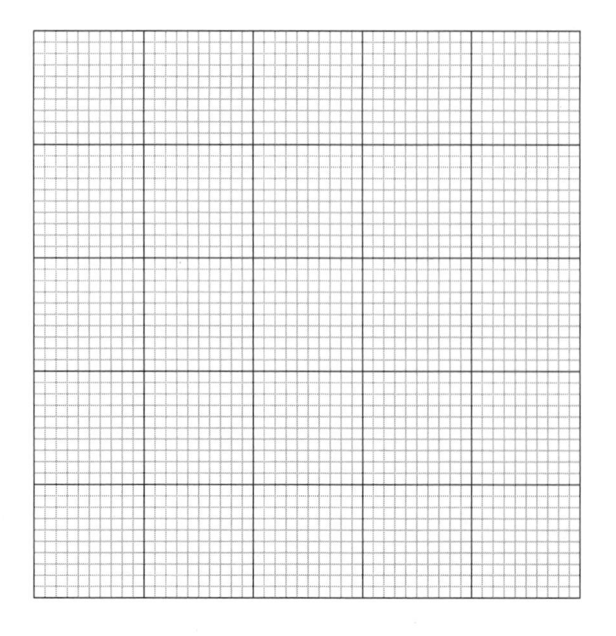

b Describe what your scatter diagram shows. (2 marks)

State whether the correlation is positive or negative. As it is worth 2 marks, describe what this means, i.e. what can you say about how the students who scored higher marks in the maths test did in the physics test?

c Circle the outlier. (1 mark)

d Give a sensible reason for the outlier. (1 mark)

[Total: 7 marks]

3 A second-hand computer shop recorded laptops' selling prices against their age.

The scatter diagram shows the results. (★★★★★)

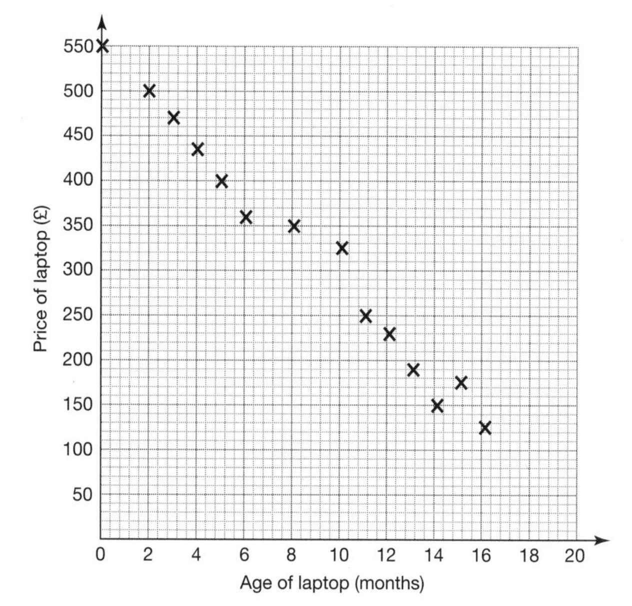

NAILIT!

When drawing a line of best fit, try to ensure the same number of crosses is on each side of the line. There may or may not be crosses on the line too.

a The shop owner tells an employee that, on average, a laptop loses £150 every 6 months.

Is he correct? Show how you get your answer. (3 marks)

Draw a line of best fit on the scatter diagram. Does it agree with what the shop owner says?

b You should not use a line of best fit to predict the price of a laptop at 20 months. Give one reason why. (3 marks)

[Total: 6 marks]

Practice paper (non-calculator)

Foundation tier

Time: 1 hour 30 minutes

The total mark for this paper is 80.
The marks for **each** question are shown in brackets.

1 Write down the value of the 7 in the number 27 892.

....

[1 mark]

2 Work out 40% of 50.

....

[1 mark]

3 Sandeep says, 'This fraction, decimal and percentage are equal.'

$\frac{2}{5}$ 0.4 4%

Is Sandeep correct?

You must show working to justify your answer.

....

[2 marks]

4 Write down the prime factors of 45.

....

[2 marks]

5 The diagram below represents two lamp posts in a street.

Lamp post A
X

Lamp post B
X

Scale: 1 cm represents 20 metres.

Work out the distance, in metres, between lamp post A and lamp post B.

....

[1 mark]

6 Larry makes this fair spinner with his name on it.

He spins the spinner once.

What is the probability that it lands on r?

L a r r y

....

[1 mark]

7 A charity wants to raise £25 000. It holds two fund-raising events: a fun run and a music festival.

The fun run raises £9689. The music festival raises £6370 more than the fun run.

Does the charity raise £25 000?

You must show working to justify your answer.

....

[3 marks]

8 A box contains dark and milk chocolates. The ratio of the number of dark chocolates in the box to the number of milk chocolates in the box is 3 : 7.

What fraction of the chocolates are milk?

....

[2 marks]

9 A cinema offers two different prizes in a competition.

Prize A: receive 8 tickets each month for four months.

Prize B: receive 2 tickets this month, then twice as many next month, twice as many again the month after that... and so on for four months.

Which prize gives more tickets?

You must show working to justify your answer.

....

[3 marks]

10 Here are the first three patterns in a sequence. The patterns are made from squares and right-angled triangles.

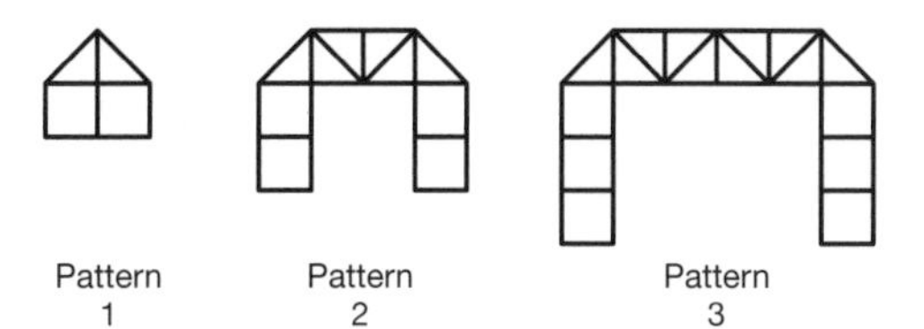

a How many right-angled triangles are there in Pattern 8?

....

[2 marks]

b Harry says there are 6 triangles in Pattern 2 so there will be 12 triangles in Pattern 4.

Is Harry correct? Give a reason for your answer.

....

....

[1 mark]

[Total: 3 marks]

11 This back-to-back stem and leaf diagram shows the cost of parking for 1 hour or all day, in different town car parks.

1 hour		all day
80 60	0	
70 50 20 20 20 00	1	
60 50 30 00 00	2	
50 00	3	00
	4	00 00 50 80
	5	00 00 00 50
	6	20 50 90
	7	50

Key	
1 hour parking	all day parking
60|0 means £0.60	3|00 means £3.00

a What is the most expensive price for parking for 1 hour?

....

[1 mark]

b What is the modal price for parking all day?

....

[1 mark]

[Total: 2 marks]

12 A plumber charges a £45 call out fee and then £35 per hour. She also charges VAT at 20% on all her bills. How much is the bill if you call her out for 3 hours' work?

....

[2 marks]

13 Buses 2A and 2B both arrive at a bus stop at 10 am.

Bus 2A stops there again every 15 minutes. Bus 2B stops there again every 12 minutes.

At what time will they both arrive at the bus stop together again?

....

[2 marks]

14 60 patients are treated by a dental practice in one day.

Children	Adults	Senior citizens
10	45	5

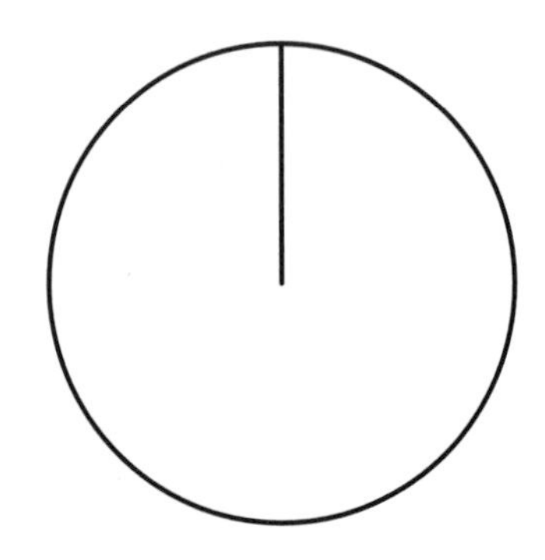

Draw an accurate pie chart for this information.

[3 marks]

15 Derek buys 8 packs of T-shirts. Each pack costs £12 and contains 6 T-shirts.
On his Saturday market stall Derek sells $\frac{3}{8}$ of the T-shirts, for £5 each. On his Sunday stall he sells the rest of the T-shirts, for £3 each.
How much profit does Derek make?

....

[4 marks]

16 a Work out $\frac{5}{8} - \frac{7}{12}$

....

[2 marks]

b Work out $2\frac{2}{3} \div \frac{4}{9}$

....

[2 marks]

[Total: 4 marks]

17 a Simplify $8 - 3x + 5 - x$

....

[2 marks]

b Expand and simplify $(x + 3)(x - 4)$

....

[2 marks]

[Total: 4 marks]

18 In a sale, a computer is reduced by 30%. This is a saving of £45.
Work out the original price of the computer.

....

[2 marks]

19 Here is a grid showing the triangle *ABC*.

a What are the coordinates of point *A*?

....

[1 mark]

b Write down the midpoint of side *BC*.

....

[2 marks]

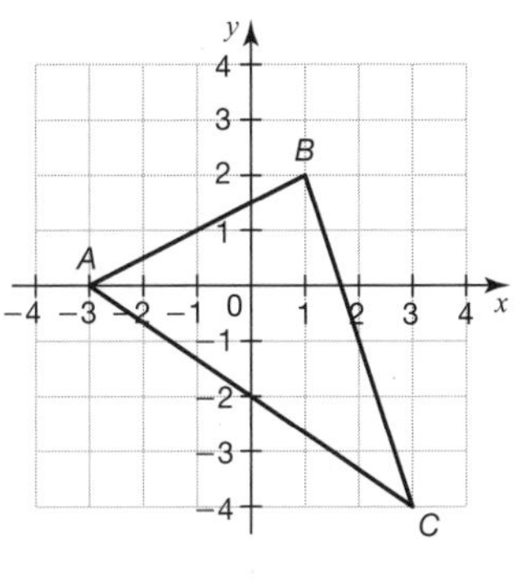

[Total: 3 marks]

20 Work out 8.15×3.4

....

[3 marks]

21 Here are sketches of the plan, front and side elevations of a 3D shape made with identical 1 cm^3 cubes.

Sketch the 3D shape.

Plan:

Front:

Side:

[2 marks]

22 Two fences meet at an angle of 65°. A straight footpath is to be installed between them, so that it is the same distance from both fences. Draw the fences and the line that the footpath should take.

[3 marks]

23 Estimate an answer to this calculation.

$$\frac{8.02 \times 3.76}{15.98}$$

....

[3 marks]

24 A science formula states.

$$K = \frac{1}{2}mv^2$$

a Find the value of K when $m = 11$ and $v = 3$.

....

[2 marks]

b Find the value of v when $K = 180$ and $m = 10$.

....

[2 marks]

[Total: 4 marks]

25

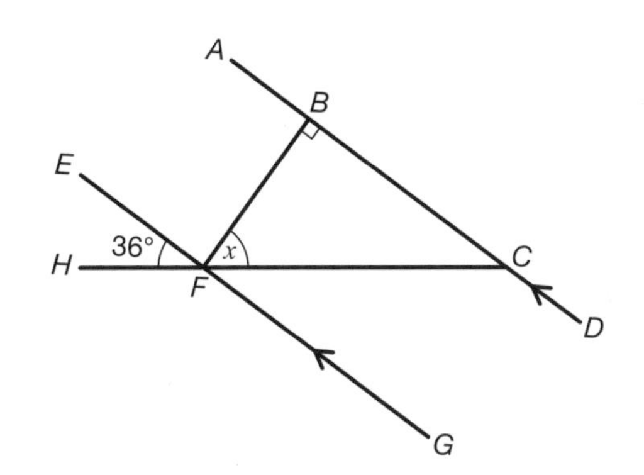

AD and *EG* are parallel.

Triangle *BCF* is a right-angled triangle.

Angle *EFH* = 36°

Work out the size of the angle marked x.

You must show your working.

....

[3 marks]

26 The graph shows a train journey.

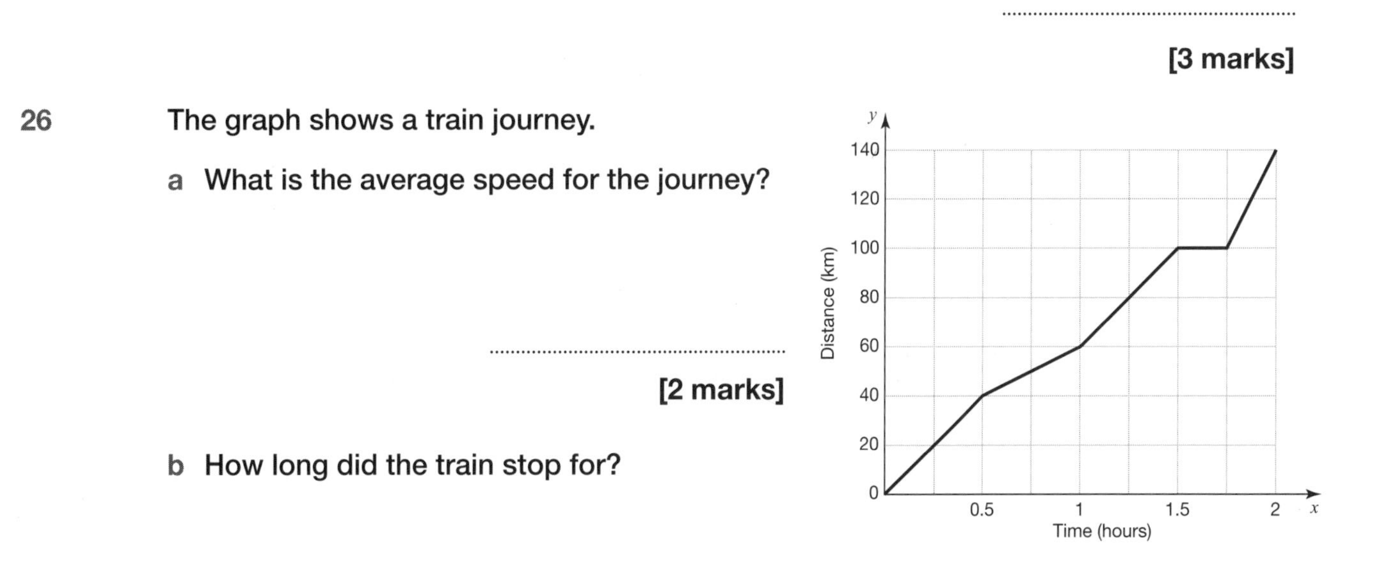

a What is the average speed for the journey?

....

[2 marks]

b How long did the train stop for?

....

[1 mark]

c What was the fastest speed the train travelled?

....

[2 marks]

[Total: 5 marks]

27 Here is an isosceles triangle.

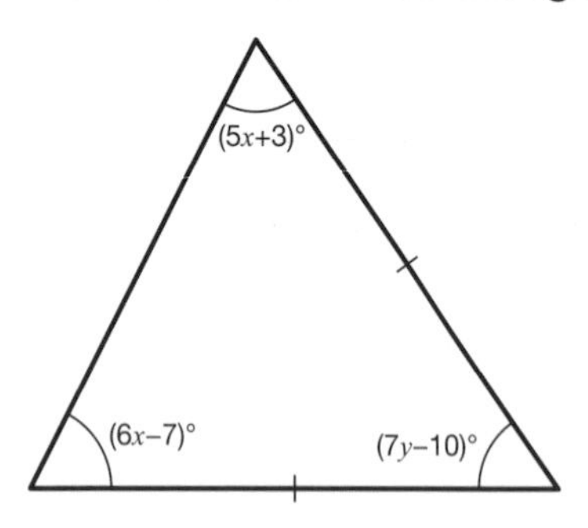

Work out the value of x and the value of y.

....

[4 marks]

28 **a** On the grid, rotate shape A 90° anti-clockwise about (0, 1). Label the new shape B.

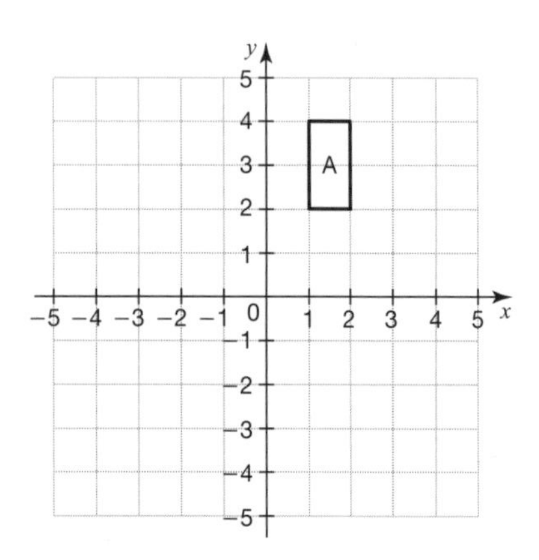

[2 marks]

b Translate shape B by vector $\begin{pmatrix} 5 \\ -1 \end{pmatrix}$. Label the new shape C. **[2 marks]**

c Describe fully the single transformation that maps shape C onto shape A.

....

[2 marks]

[Total: 6 marks]

29 $\mathbf{a} = \begin{pmatrix} -1 \\ 2 \end{pmatrix}$

$\mathbf{b} = \begin{pmatrix} 7 \\ 7 \end{pmatrix}$

Work out **b** − 3**a** as a column vector.

....

[2 marks]

[Total marks: 80]

Practice paper (calculator)

Foundation tier

Time: 1 hour 30 minutes

The total mark for this paper is 80.
The marks for **each** question are shown in brackets.

1 Write $\frac{7}{8}$ as a decimal.

....

[1 mark]

2 Work out 140% of £350.

....

[1 mark]

3 Eliza gets £83.75 for working 5 hours. How much will she get for working 37 hours?

....

[2 marks]

4 A games console costs £255 in the UK. The same one costs 300 euros in France.
The exchange rate is £1 = €1.16.
Is it cheaper to buy the games console in the UK or France?
Show working to justify your answer.

....

[3 marks]

5 Find the value of $\sqrt[3]{13.824} + (4.5 - 0.38)^2$.
Give your answer to 3 significant figures.

....

[3 marks]

6 XYZ is a right-angled triangle.
Work out the length of XY.
Give your answer correct to 1 decimal place.

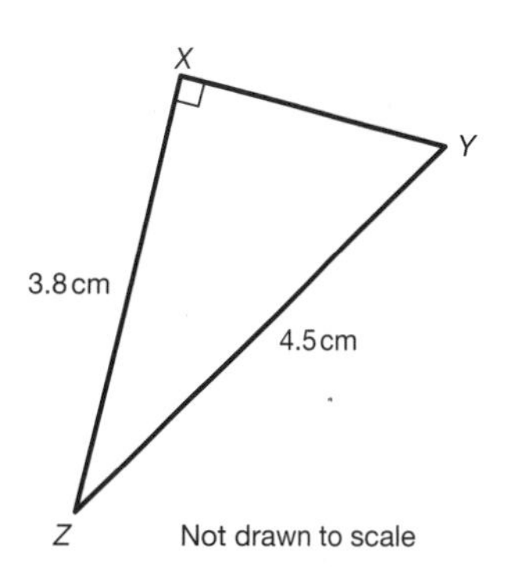

....

[3 marks]

7 The bar chart shows the number of work and personal emails Jeff received in one working week.

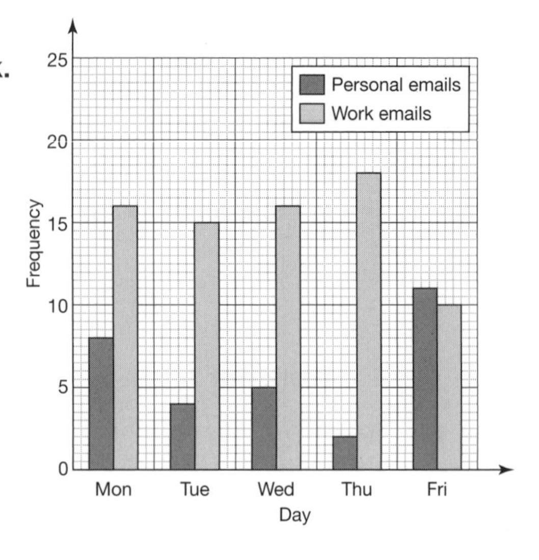

a On which day did he receive more personal than work emails?

....

[1 mark]

b What percentage of the emails he received on Thursday were personal?

....

[2 marks]

c What fraction of the emails he received in the week were work emails?

....

[2 marks]

[Total: 5 marks]

8 Andrew writes down a factor of 12 and a multiple of 9.

He multiplies his two numbers together. His answer is greater than 80 but less than 90.

Find the two pairs of numbers Andrew could have written down.

....

[3 marks]

9 Write down the integer values of x that satisfy the inequality $-1 \leq x < 5$.

....

[2 marks]

10 A 5-litre tin of fence paint costs £22.50. The paint tin says the paint will cover $20\,\text{m}^2$.

Geraldine has 16 fence panels to paint, each measuring $1.8\,\text{m} \times 1.8\,\text{m}$.

How much will Geraldine spend on fence paint?

Show working to justify your answer.

....

[3 marks]

11 This is a box made to look like a house.

It is to be filled with sweets. What volume of sweets will it hold?

8 cm
9 cm
7 cm
6 cm

....

[2 marks]

12 Change 84 km/h into m/s.

....

[3 marks]

13 Jessie and Lucy share some money in the ratio 2 : 3. Lucy receives £30 more than Jessie. How much money does Jessie get?

....

[3 marks]

14 Points *P* and *Q* are shown on a kilometre square grid. A ship sails from point *P* to point *Q*. Work out the distance the ship sails. Give your answer as a surd.

y
5
4
3
2
1
0
1 2 3 4 5 x
P
Q

....

[3 marks]

15 At a circus, 50% of the audience are children.

Of the remaining audience, $\frac{1}{3}$ are men and the rest are women. There are 42 women.

How many people in total are in the audience at the circus?

....

[3 marks]

16 The distance between the Sun and the planet Mercury is approximately 58 million km. Write this distance in standard form.

....

[2 marks]

17 Tom, Rohan and Seren throw balls into a basketball hoop. Tom gets 3 more balls in the net than Seren. Rohan gets 2 more balls in the net than Tom. Write an expression for the total number of balls they all get in the net.

....

[3 marks]

18 Sam gets the train to work. On each day the probability the train is late is $\frac{1}{8}$.

a Complete the tree diagram.

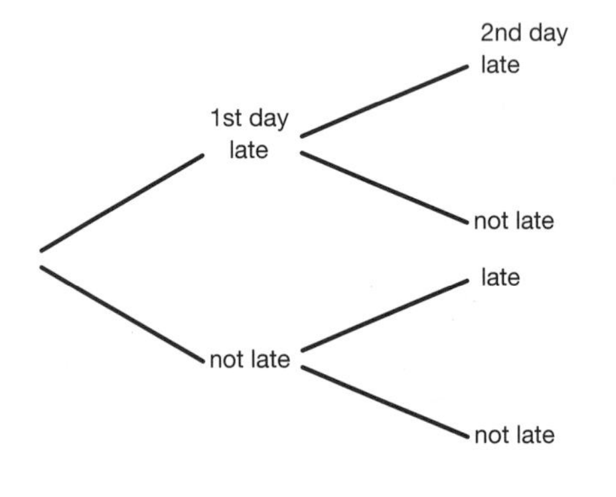

[3 marks]

b Sam only works two days in Easter week. Work out the probability that the train is late on one day and not the other.

....

[3 marks]

[Total: 6 marks]

19 Triangles *ABC* and *BCD* are right-angled triangles.

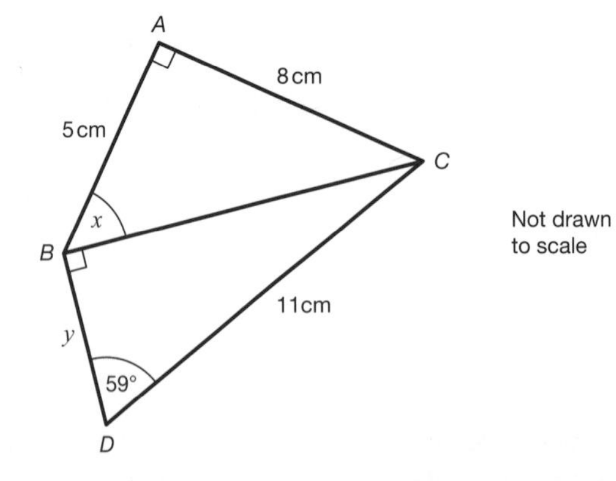

Work out angle x to the nearest degree and length y to 2 decimal places.

....

[6 marks]

20 The table shows the goals scored by two different football teams during the same tournament.

Team A	
Goals	**Frequency**
0	3
1	4
2	4
3	2
4	1
5	2

Team B	
Goals	**Frequency**
0	3
1	4
2	8
3	4
4	1

Which team scored more goals, on average?

....

[3 marks]

21 The graph shows two different ways to pay for gym membership each month.

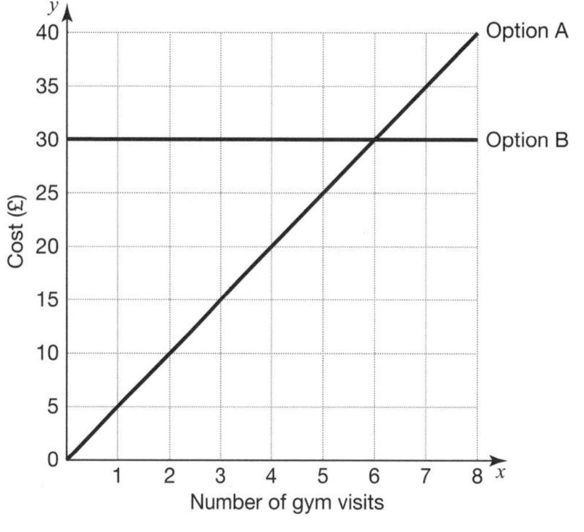

a Explain which option is better value if you plan to visit the gym once per week.

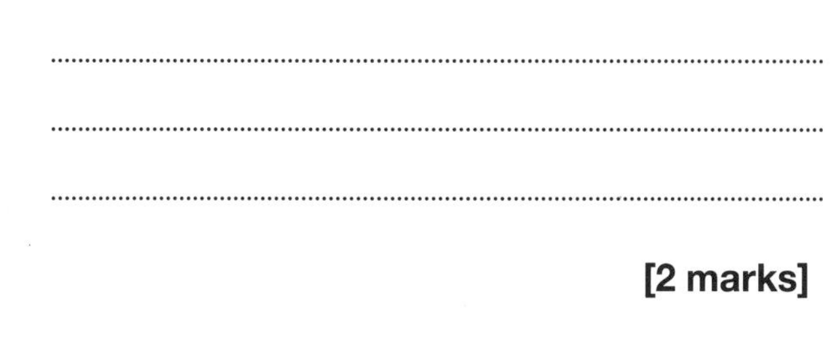

....

....

....

[2 marks]

b After how many visits to the gym each week does it make sense to pay a monthly fee?

....

[2 marks]

[Total: 4 marks]

22 A semicircular lawn has three identical flowerbeds that are smaller semicircles. The large semicircle has diameter *AB* of 15 m. The centre of each of the smaller semicircles lies on *AB*.

Work out the area of the lawn. Give your answer to 3 significant figures.

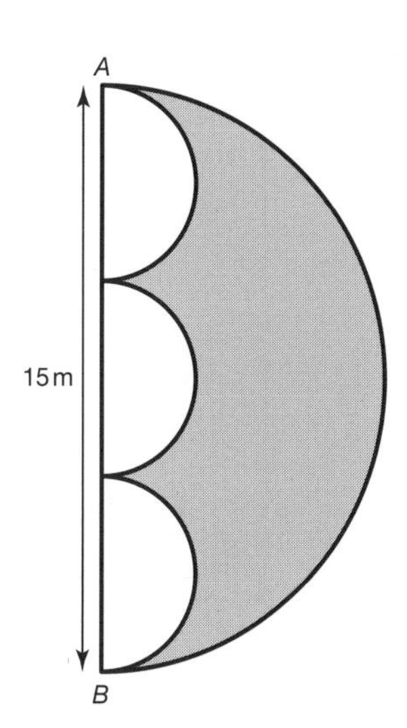

....

....

[3 marks]

23 a Complete the table of values for $y = 2x^2 - 3$.

x	−2	−1	0	1	2
y					

[2 marks]

b On the grid below, draw the graph of $y = 2x^2 - 3$ for values of x from $x = -2$ to $x = 2$.

[2 marks]

[Total: 4 marks]

24 A car dealer records the price and mileage of cars he sells. The scatter diagram shows the information.

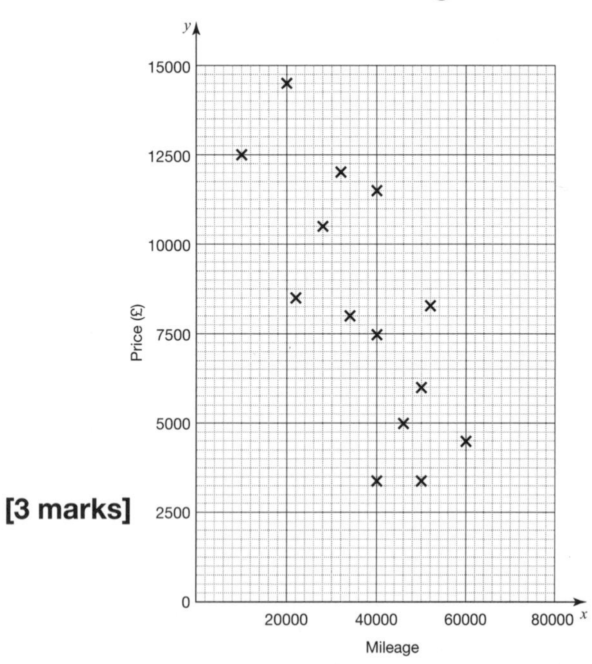

The car dealer says, 'A car depreciates by approximately £2500 with each 10000 miles.'

a Is the car dealer correct? Show how you get your answer.

[3 marks]

b How much should the car dealer price a car with a mileage of 55000 miles?

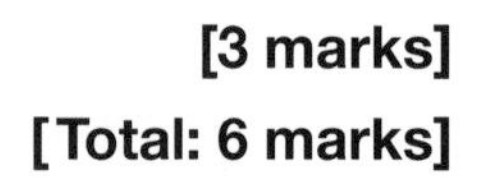

[3 marks]

[Total: 6 marks]

25 The length of a rectangle is $n + 5$. Its width is $n - 3$.

Write the area of the rectangle in its simplest form.

....

[3 marks]

[Total marks: 80]

Answers

For full worked solutions, visit:
www.scholastic.co.uk/gcse

Number

Factors, multiples and primes

1 6
2 17, 19, 23
3 $60 = 2^2 \times 3 \times 5$
4 Drummer 1 hits her drum at: 6 12 18 24 30 36 42 48 54 60 seconds
Drummer 2 hits his drum at: 8 16 24 32 40 48 56 seconds
They hit their drums at the same time twice (two times) after 24 and 48 seconds.

Ordering integers and decimals

1 −12, −8, −1, 0, 2
2 0.32, 0.3, 0.23, 0.203
3 **a** $-4 < 0.4$ **c** $-0.404 > -0.44$
b $4.200 < 4.3$ **d** $0.33 < 0.4$

Calculating with negative numbers

1 **a** −10 **b** −4 **c** 5 **d** 1
2 **a** −18 **b** 4 **c** 40 **d** −16
3 −7° C
4 1 correct answer; 4 incorrect answers

Multiplication and division

1 **a** 2142 **b** 11 223 **c** 92 **d** 52
2 **a** 12 **b** 12
3 £335
4 1196 hours

Calculating with decimals

1 76.36
2 £7.51
3 38.29
4 Flo raises £28.75; Kirsty raises £143.75

Rounding and estimation

1 **a** 0.798 **b** 0.80
2 5
3 **a** £7500
b Overestimate, because the concert ticket price and number of tickets sold were rounded up, and so the amount of income was estimated more than it really is.

Converting between fractions, decimals and percentages

1 **a** $\frac{71}{1000}$ **c** 40%
b 0.63 **d** $\frac{8}{25}$
2 **a** 0.3125 **b** 31.25%
3 $\frac{5}{8} = 0.625$ 0.65 60% = 0.6
Therefore, 0.65 is largest.

Ordering fractions, decimals and percentages

1 **a** $\frac{1}{2} < 0.6$ **b** $\frac{3}{4} > 0.7$ **c** $-\frac{3}{10} < 0.2$
2 **a** $\frac{5}{12}$ $\frac{9}{20}$ $\frac{7}{15}$
b $\frac{1}{25}$ 0.4 45%
3 $\frac{1}{3} = 33.3\%$; $\frac{2}{5} = 40\%$, so shop C, shop A, shop B
4 $\frac{5}{9}$ 38.5% 0.38 $\frac{3}{10}$

Calculating with fractions

1 $\frac{29}{45}$ 2 $\frac{1}{12}$ 3 $\frac{11}{21}$ 4 10

Percentages

1 10
2 £13.60
3 14 193
4 £1008

Order of operations

1 7 2 23 3 4.0964

Exact solutions

1 $0.133\,\text{cm}^2$
2 $1\frac{7}{9}\,\text{m}^2$
3 $2\sqrt{3}\,\text{cm}^2$
4 Area of a circle $= \pi r^2$
The fraction of the circle shown $= \frac{3}{4}$
The area of the circle shown $= \frac{3}{4} \times \pi r^2$
Radius = 2 cm
The area of the shape $= \frac{3}{4} \times \pi \times 2^2 = \frac{3}{4} \times \pi \times 4 = 3\pi$

Indices and roots

1 **a** 7^4 **b** 5^{-3}
2 **a** 16 **b** $\frac{1}{100}$
3 $3^{-2} = \frac{1}{9}$ $\sqrt[3]{27} = 3$ $\sqrt{25} = 5$ $2^3 = 8$
4 1

Standard form

1 2750 3 6.42×10^{-3}
2 1.5×10^8 4 2.8×10^{-4} km

Listing strategies

1 259, 295, 529, 592, 925, 952
2 **a**

		4-sided spinner			
		0	1	2	3
3-sided spinner	1	1	2	3	4
	2	2	3	4	5
	3	3	4	5	6

b 4
3

		Dice					
		1	2	3	4	5	6
Coin	H	H1	H2	H3	H4	H5	H6
	T	T1	T2	T3	T4	T5	T6

4 spj; spi; sfj; sfi; bpj; bpi; bfj; bfi

Algebra

Understanding expressions, equations, formulae and identities

1 **a** identity **b** equation **c** expression
2 **a** equation, because it has an equals sign and can be solved
b formula, because it has letter terms, an equals sign and the values of the letters can vary
c an expression because it has letter terms and no equals sign
d formula, because it has letter terms, an equals sign and the values of the letters can vary
3 **a** Any of: $2x + 10$ or $10x + 2$ or $x + 210$ or $x + 102$
b Any of: $2x = 10$ or $10x = 2$

Simplifying expressions

1 $8x$
2 **a** $48a^2$ **b** $30p^3$
3 $5y$
4 $8u$

Collecting like terms

1 a $3m + 4n$ b $-3q - 2r$
2 a $a + 13b$ b $-c + d$
3 a $5p^3 + p$ b $-7x^2 + 3x + 12$
4 $-5\sqrt{5} + f$

Using indices

1 a p^4 b $12y^5$ c $10a^5b^3$
2 a q^{-6} b u^{-6} c 1
3 a b b f^3 c $\frac{y^2}{x}$
4 $(2m^3)^3$

Expanding brackets

1 a $4m + 12$ b $2p - 2$ c $30x - 50$
2 a $8m + 11$ b $4x + 2$
3 a $y^2 + 10y + 21$
b $b^2 - 2b - 8$
c $x^2 - 10x + 24$
4 a $q^2 + 2q + 1$ b $z^2 + 4z + 4$ c $c^2 - 6c + 9$

Factorising

1 a $4(x + 2)$ b $3(d - 5)$ c $4(2y - 3)$
2 a $q(q + 1)$ b $a(a + 6)$ c $5z(2z + 3)$
3 a $(x + 3)(x + 4)$
b $(x - 2)(x + 8)$
c $(a - 6)(a - 4)$
4 a $(y + 2)(y - 2)$
b $(x + 3)(x - 3)$
c $(p + 10)(p - 10)$

Substituting into expressions

1 2
2 44
3 a −3 b −28 c 8 or −8

Writing expressions

1 a $n + 3$ b $2n - 9$
2 a $x + y$ b $5x$ c $12x + 11y$
3 $28p + 4$
4 $s(5s + 1)$

Solving linear equations

1 a $x = 7$ b $x = 13$ c $x = 5$ d $x = 18$
2 a $x = 6$ b $x = 7$ c $x = 25$ d $x = 3$
3 a $x = 1$ b $p = 4$ c $m = 2$ d $q = 4$
4 a $x = 5$ b $y = 4$ c $x = 3$ d $n = 6$

Writing linear equations

1 $x = 20$
2 12 years old
3 130 cm²
4 $x = 15, y = 4$

Linear inequalities

1 a −1, 0, 1, 2, 3, 4, 5
b −3 −2 −1 0 1 2 3 4 5 6
2 a $x > 5$
4 5 6 7 8 9 10
b $x \leq 7$
4 5 6 7 8 9 10
c $x < 8$
4 5 6 7 8 9 10
3 a $-1 \leq x < 2$
b $-2 < x \leq 0$
4 1, 2, 3, 4, 5

Formulae

1 a 1 hour 40 minutes b 12.10 pm
2 a $C = l + nk$ b £109.50
3 a $q = \frac{3p}{s}$ c $q = \frac{p - 3r}{3}$
b $q = rp - rt$ d $q = \frac{p^2}{2}$

Linear sequences

1 a 6, 11, 16, 21 b 251
2 a 19
b No, Rachel is not correct, because the number of triangles is not the pattern number multiplied by 2. Instead, it is the pattern number add 2, so there will be 6 triangles in pattern 4.
3 a $11n - 8$
b No, because then $11n - 8 = 100$ so $11n = 108$, n is not a whole number.

Non-linear sequences

1 a 16, 32 b 0.1, 0.01 c 24, −48 d geometric
2 a 15 b 9th
3 a Yes, he is correct.

Day	Mon	Tue	Wed	Thu	Fri
Number of ladybirds	2	8	32	128	512

b Saturday
4 a $\frac{1}{2}, 2, 4\frac{1}{2}$
b Yes, because $\frac{1}{2}n^2 = 32$, $n^2 = 64$, $n = 8$

Show that...

1 LHS = $2x + 1$; RHS = $2x + 1$; LHS = RHS. Therefore, $2\left(x + \frac{1}{2}\right) \equiv x + x + 1$
2 LHS = $x^2 - 25 + 9 = x^2 - 16$; RHS = $x^2 - 16$
3 Let the three consecutive numbers be n, $n + 1$ and $n + 2$. $n + n + 1 + n + 2 = 3n + 3 = 3(n+1)$. Therefore, the sum of three consecutive numbers is a multiple of 3.
4 a width of pond = $x - y + x + x + x - y = 4x - 2y$
length of pond = $4x$
Perimeter = $4x - 2y + 4x - 2y + 4x + 4x = 16x - 4y$
b Yes Sanjit is correct, because $16x - 4y = 4(4x - y)$, showing that when x and y are whole numbers, the perimeter is always a multiple of 4.

Functions

1 a 11 b 6 c $y = 4x - 1$

2

x	y
−2	−1
0	3
3	9

3

x	y
−2	0
1	$1\frac{1}{2}$
8	5

Coordinates and midpoints

1 a (1, 2)
b
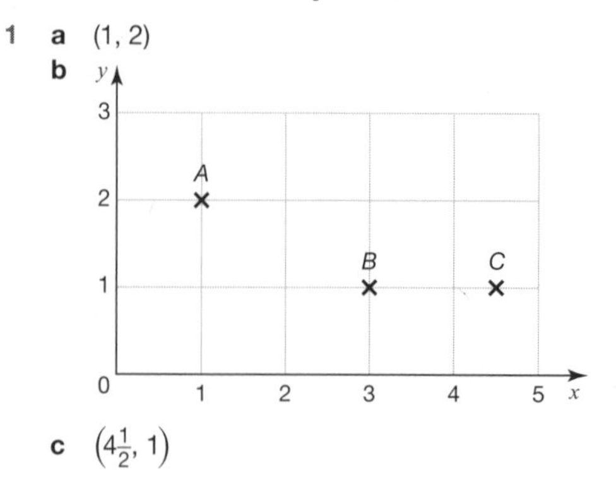

c $\left(4\frac{1}{2}, 1\right)$

2 **a** (1, −4)

b

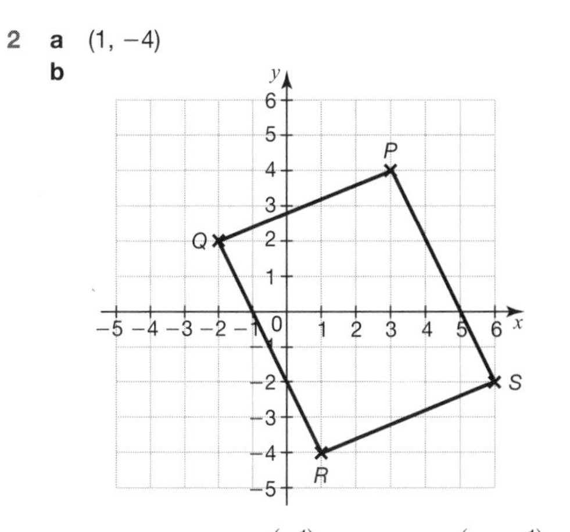

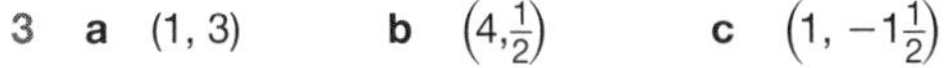

3 **a** (1, 3) **b** $\left(4, \frac{1}{2}\right)$ **c** $\left(1, -1\frac{1}{2}\right)$

Straight-line graphs

1 **a**

x	−1	0	1	2
y	1	3	5	7

b

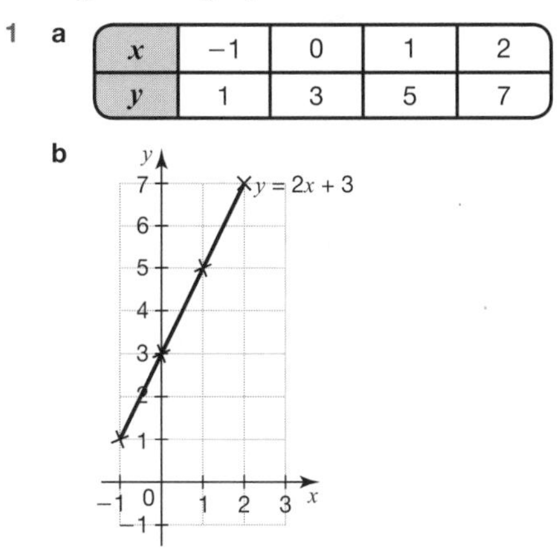

2 **a** and **b**

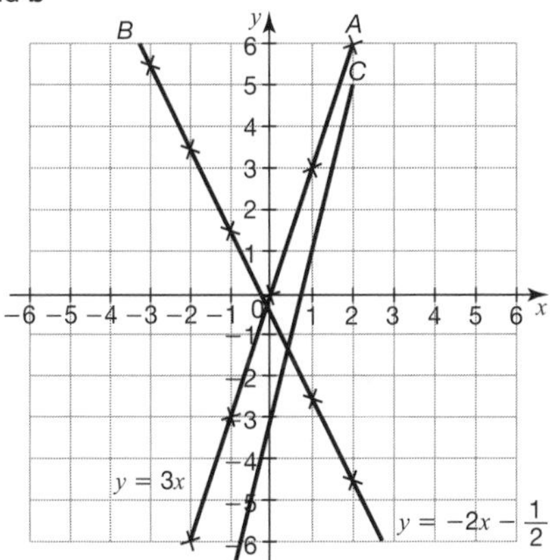

c $y = 4x - 3$

3 **a** B and C, because they have the same gradient of 2.

b A and B, because they both have a y-intercept at (0, 1).

4 $y = 2x - 4$

5 **a** 2 **b** 1

6 **a** 2.5 **b** 4 **c** 0.6

7 **a**

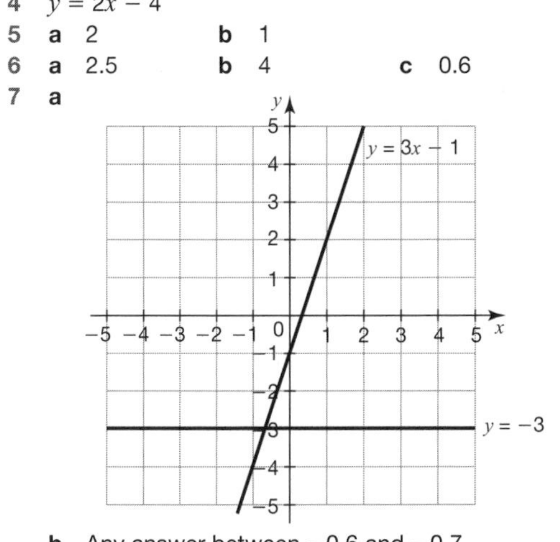

b Any answer between −0.6 and −0.7.

8

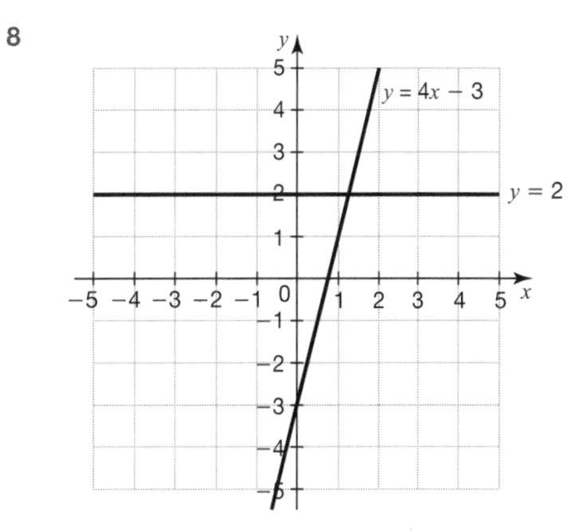

Any answer between 1.2 and 1.3.

Solving simultaneous equations

1 **a** $x = 3, y = 6$

b $x = 4, y = 1$

c $x = 2, y = -2$

2 **a** $x = 2, y = 5$

b $x = -1, y = -3$

c $x = 1, y = 6$

3 **a** $x + y = 2$ $2x - y = 1$

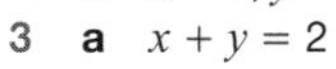

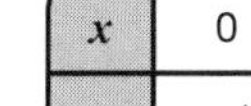

x	0	2
y	2	0

x	0	$\frac{1}{2}$
y	−1	0

b

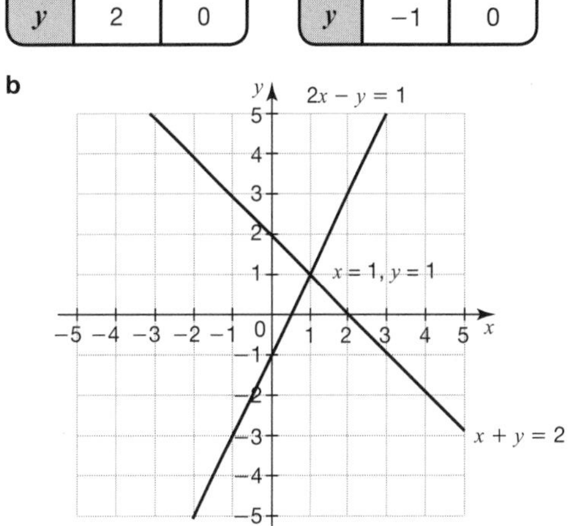

c

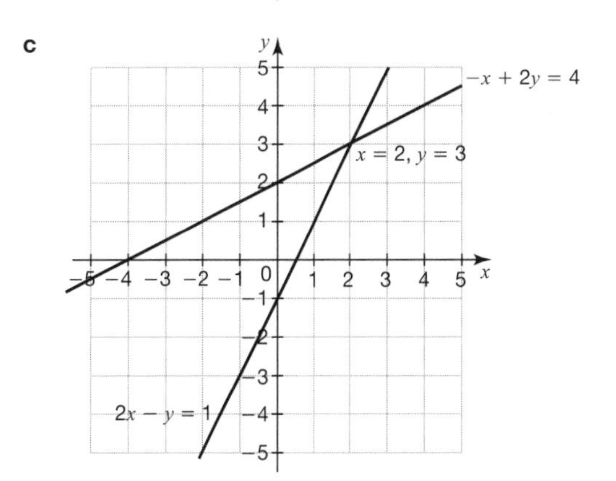

Quadratic graphs

1 **a** C and D **b** E **c** A **d** D **e** B

2 **a**

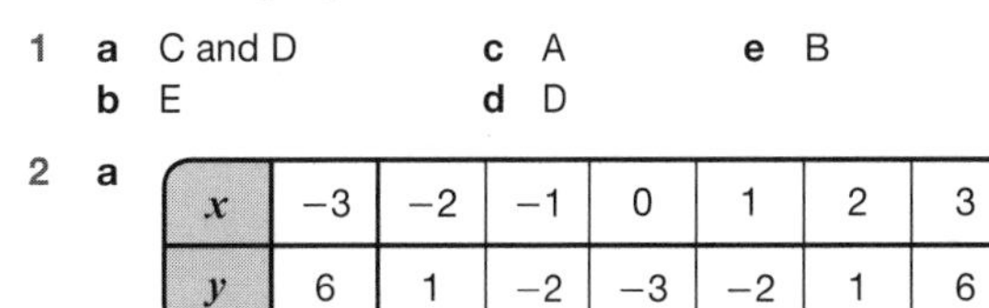

x	−3	−2	−1	0	1	2	3
y	6	1	−2	−3	−2	1	6

b

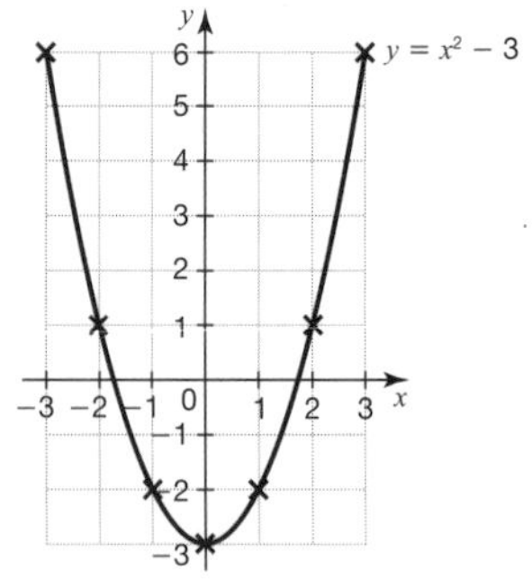

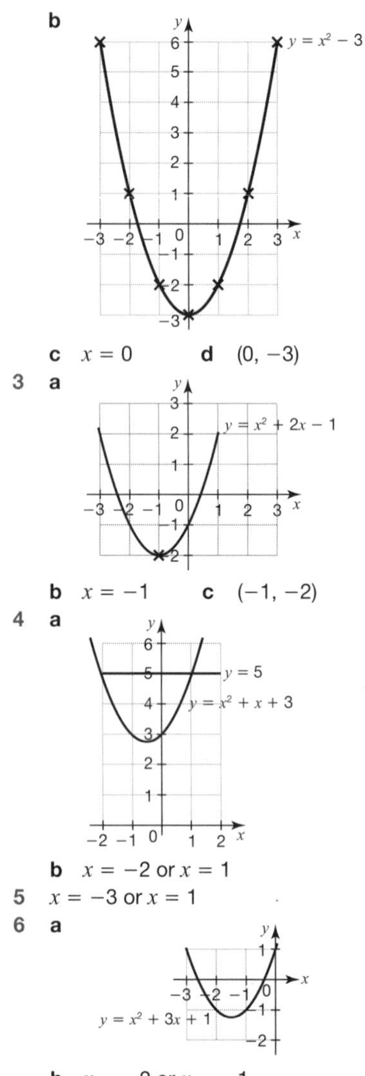

c $x = 0$ d (0, −3)

3 a

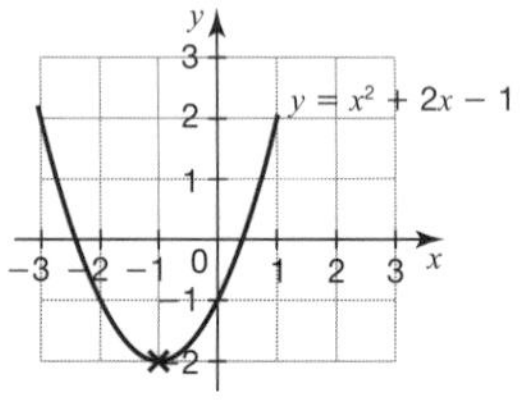

b $x = -1$ c (−1, −2)

4 a

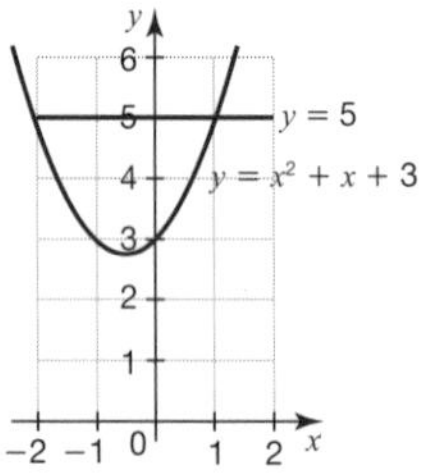

b $x = -2$ or $x = 1$

5 $x = -3$ or $x = 1$

6 a

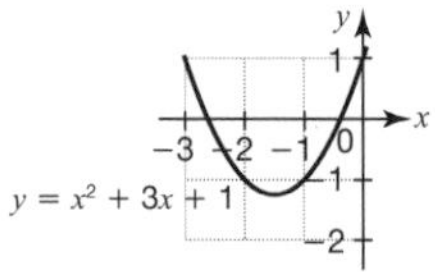

b $x = -2$ or $x = -1$

c Any answer close to $x = -2.6$ or $x = -0.38$.

Solving quadratic equations

1	a	0 or −6	b	0 or 11	c	0 or 3		
2	a	4 or −4	b	9 or −9	c	10 or −10		
3	a	−2 or −3	b	−5 or 2	c	2 or 7		
4	a	0 or 3	b	−5 or 5	c	−6 or 3		

Cubic and reciprocal graphs

1 a A, C and D b B c E d D

2 a

x	−2	−1	0	1	2
y	−4	3	4	5	12

b

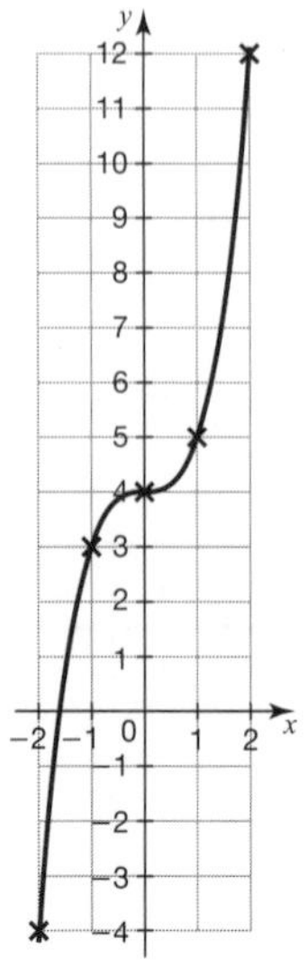

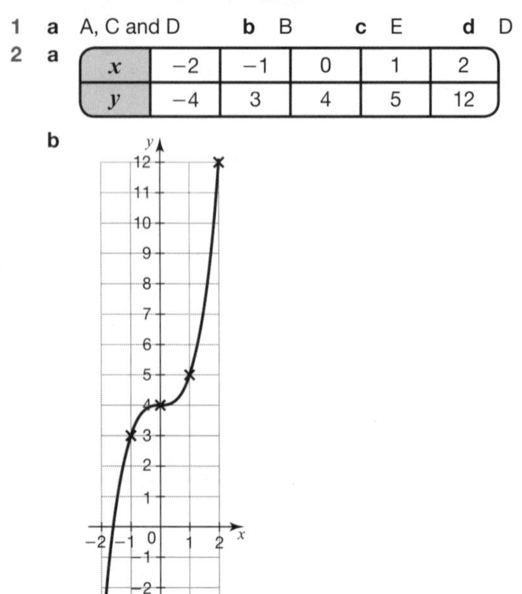

3 a cubic b (0, −8) c (2, 0)

4 a

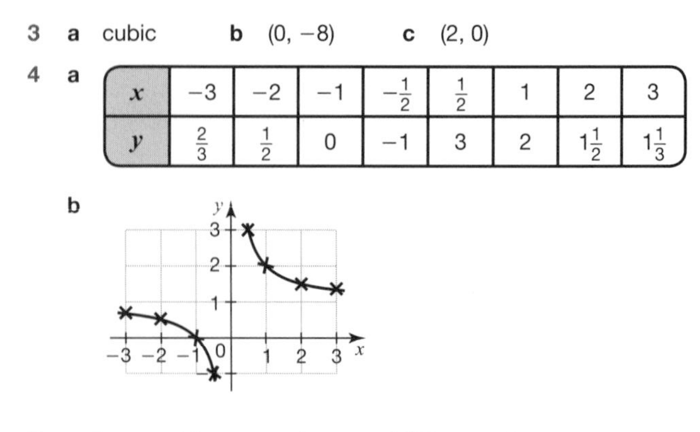

x	−3	−2	−1	$-\frac{1}{2}$	$\frac{1}{2}$	1	2	3
y	$\frac{2}{3}$	$\frac{1}{2}$	0	−1	3	2	$1\frac{1}{2}$	$1\frac{1}{3}$

b

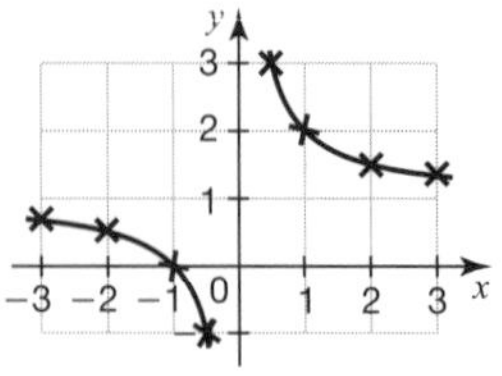

Drawing and interpreting real-life graphs

1 a \$3

b \$2.50

c

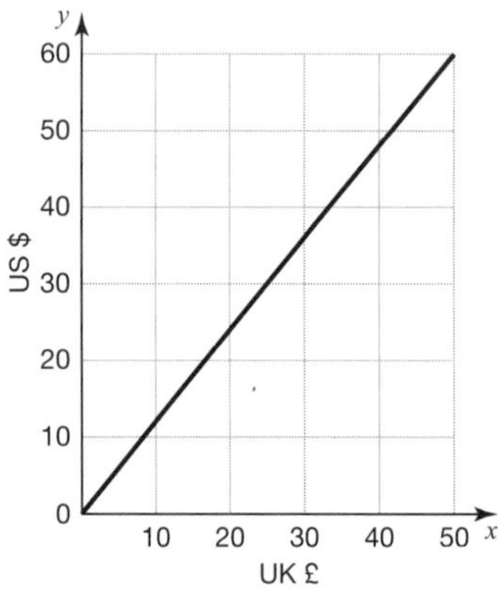

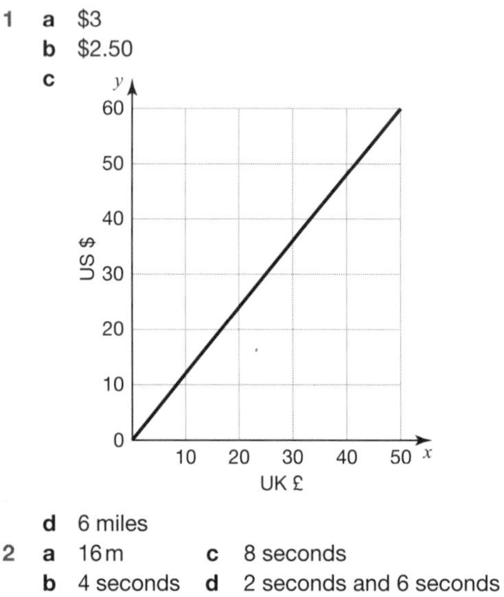

d 6 miles

2 a 16 m c 8 seconds

b 4 seconds d 2 seconds and 6 seconds

3 a 10 m/s

b The cyclist is travelling at a constant speed of 10 m/s.

c $-\frac{1}{2}$ m/s^2

d The cyclist stops

4 a A2; B4; C3; D1

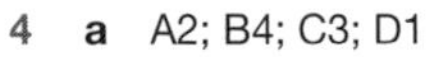

b

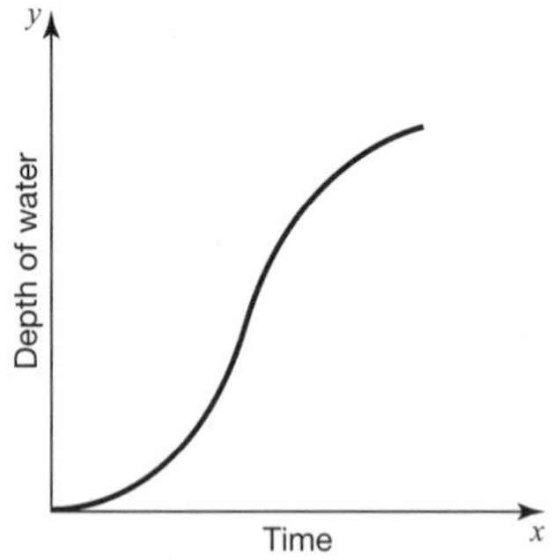

Ratio, proportion and rates of change

Units of measure

1 a 400 cm b 5 kg c 1500 ml d 8.25 km

2 2500 ml

3 a Luke: 240 seconds; Adam: 3 × 60 + 47 = 227 seconds. Adam arrived first.
or Luke: 240 ÷ 60 = 4 minutes; Adam: 3 minutes 47 seconds. Adam arrived first.

b 13 seconds

4 Ben = 1.25 m = 3.2 + 0.8 feet = 4 feet. Tom is taller.
Or Tom = 4.8 feet = 3.2 + 1.6 feet = 1 + 0.5 metres = 1.5 metres. Tom is taller.

Ratio

1 3:2 2 £10 3 £30

4

blue	blue	blue	yellow	yellow	yellow	yellow	yellow	yellow	yellow

Phil needs 3×500 ml = 1500 ml = 1.5 litres of blue paint.
He has 2 litres of blue paint.

Phil needs 7×500 ml = 3500 ml = 3.5 litres of yellow paint.
He has 3 litres of yellow paint.

Phil has enough blue paint, but does not have enough yellow paint.

Scale diagrams and maps

1 10 000 cm; 100 m
2 3 cm
3 20 m
4 24 cm

Fractions, percentages and proportion

1 $\frac{3}{4}$
2 a 1 : 3 : 6
  b $\frac{3}{10}$
  c 60%
3 $\frac{1}{80}$
4 32%

Direct proportion

1 a £16
  b £144
2 a £1.20
  b 2p
  c It is a straight-line graph; the graph passes through the origin (0, 0).
3 a 7 teaspoons of turmeric; 14 teaspoons of chilli powder; $17\frac{1}{2}$ teaspoons of cumin
  b Sally has 75 g of chilli powder. That is $75 \div 3 = 25$ teaspoons.
  Sally needs 14 teaspoons to make the curry for her class. She does have enough.

Inverse proportion

1 a £400
  b 9
  c £2000
  d Prize money and number of winners are in inverse proportion.
2 a 6 hours
  b 30 minutes (or 0.5 hours)
3 10 minutes

Working with percentages

1 40%
2 a 60%
  b 28 (£ millions)
3 a £280.90
  b £281.22

Compound units

1 50 m/s
2 4 minutes
3 6000 Newtons/m^2
4 On Saturday Sami drove $4 \times 50 = 200$ miles; on Sunday Sami drove $356 \div 8 \times 5 = 222.5$ miles. Sami drives further on Sunday.

Geometry and measures

Measuring and drawing angles

1 a 123°
  b 42°
  c 331°
2 x
3 a 100°, 120°, 140°, 160°
  b first angle + second angle = 87°. This means both angles are less than 87°, and so they both must be acute.

Using the properties of angles

1 a Angle ACB = 52° (Angles in a triangle add up to 180°)
  x = 128° (Angles on a straight line add up to 180°)
  b Angle ADC = 86° (Angles in a quadrilateral add up to 360°)
  x = 94° (Angles on a straight line add up to 180°)
2 a Angle BED = 39° (Alternate angles are equal)
  Angle BDE = 39° (Base angles in an isosceles triangle are equal)
  x = 102° (Angles in a triangle add up to 180°)
  b Angle DCF = 98° (Vertically opposite angles are equal)
  x = 98° (Corresponding angles are equal)
3 Angle CFG = 62° (Co-interior angles add up to 180°)
  x = 66° (Angles on a straight line add up to 180°)
4 $x + 40 + 3x + 5x - 40 = 180°$
  $9x = 180°$
  $x = 20°$
  Angle $BAC = x + 40 = 20 + 40 = 60°$
  Angle $ACB = 3x = 3 \times 20 = 60°$
  Angle $ABC = 5x - 40 = 5 \times 20 - 40 = 60°$
  Triangle ABC has equal angles of 60°. Therefore, it is an equilateral triangle.

Using the properties of polygons

1 a 720°
  b 120°
  c 60°
2 a octagon
  b all angles are equal; all sides are equal
  c 135°
3 exterior angle = 180° − 144° = 36°
  number of sides = 360° ÷ 36° = 10
  therefore it is a decagon.

Using bearings

1 a 065°
  b 318°
  c 245°
2 096°
3 a 060° (any value 058° to 062° accepted)
  b

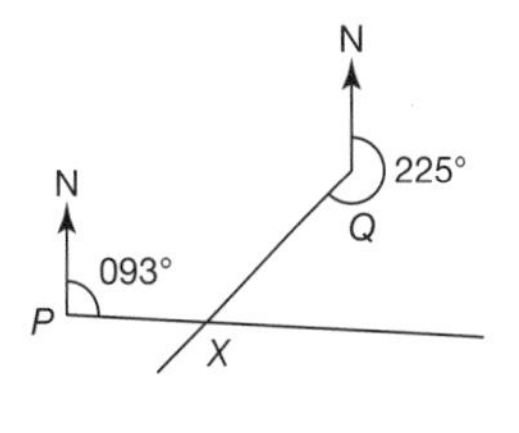

Properties of 2D shapes

1 **a**

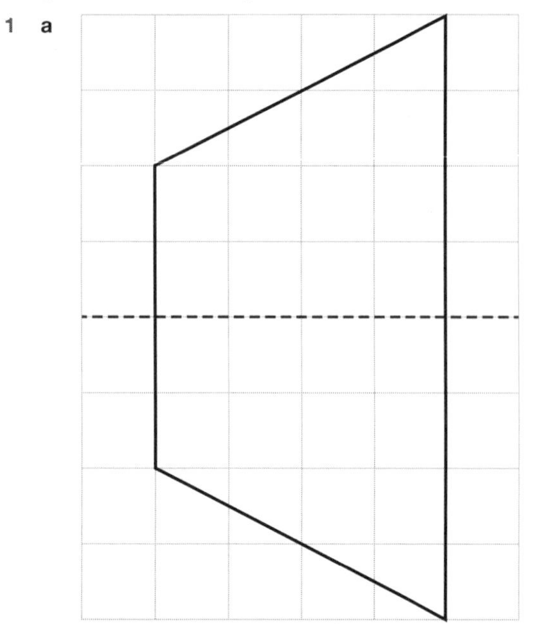

b trapezium

c one pair of parallel sides

2 **a**

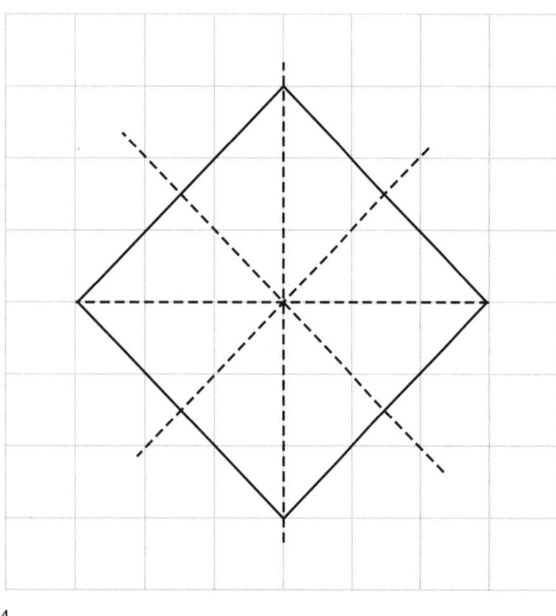

b 4

c square

d Two from: all sides equal in length; all angles are 90°; diagonals are equal; diagonals bisect each other at 90°

3 **a** rectangle, rhombus

b

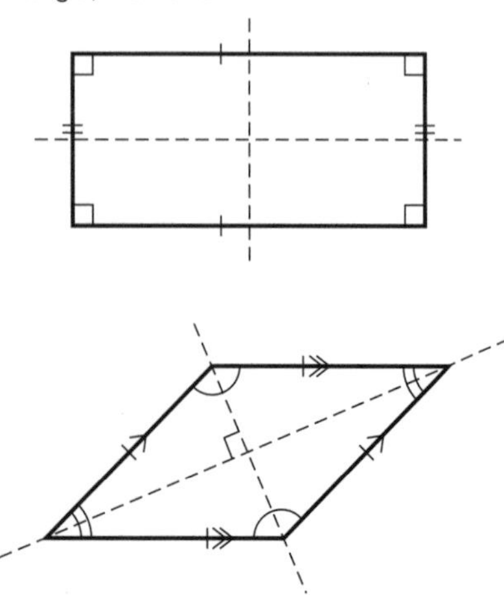

Congruent shapes

1 D and F

2 63°

3 **a** SAS

b ASA

4 No, they are not congruent. They have the same angles, but the sides may not be the same size (one triangle could be an enlargement of the other).

Constructions

1 **a** and **b**

2 **a** and **b**

3 **a**

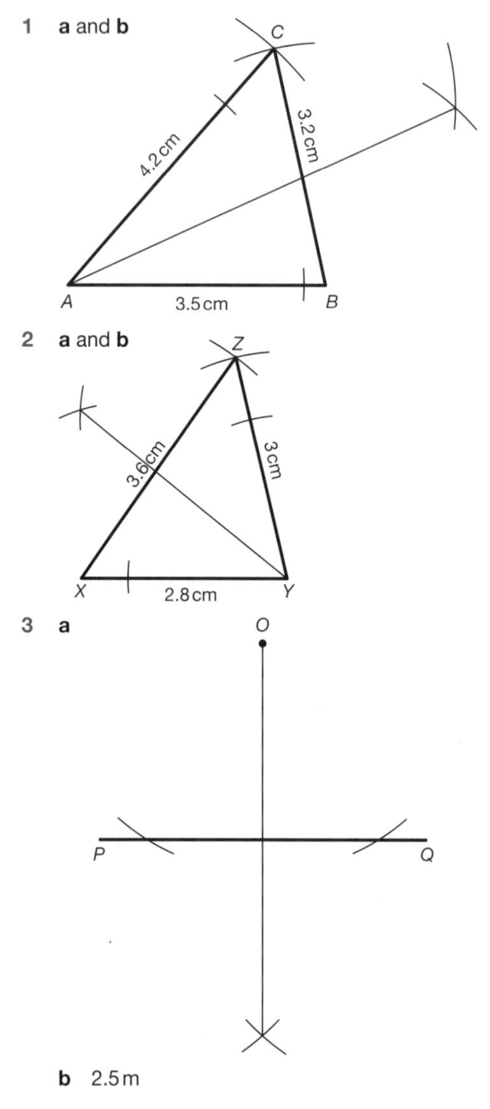

b 2.5m

Drawing circles and parts of circles

1 **a–d**

2 **a** and **b**

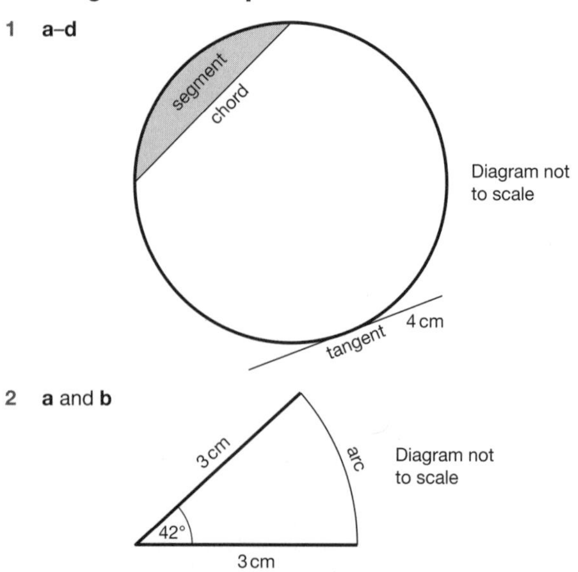

3 No, Donald is not correct. A segment of a circle is the area enclosed by a chord and an arc; a sector of a circle is the area enclosed by two radii and the arc between them.

Loci

1 a

b

c

2

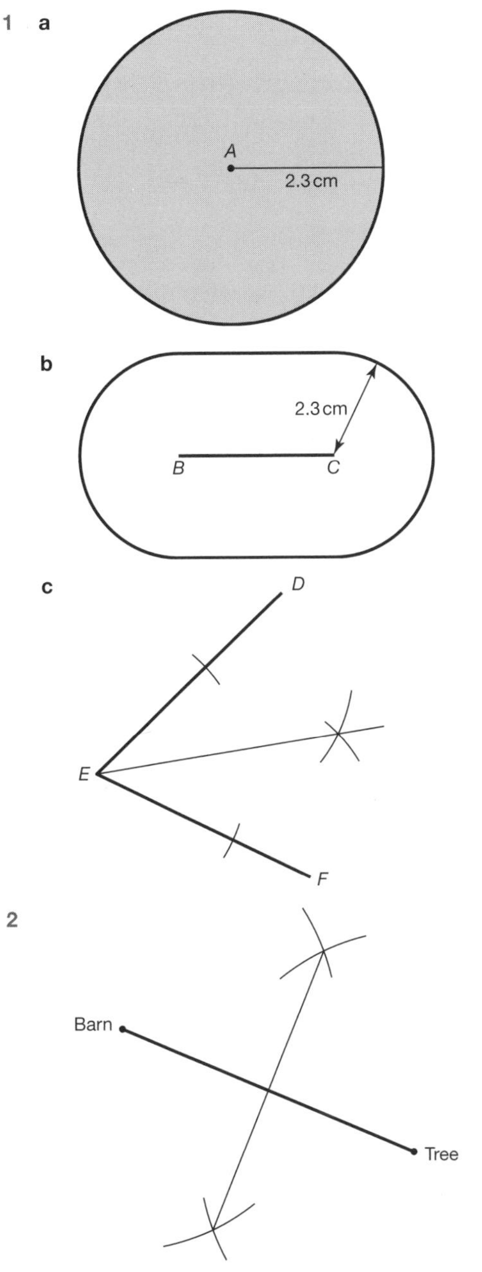

3

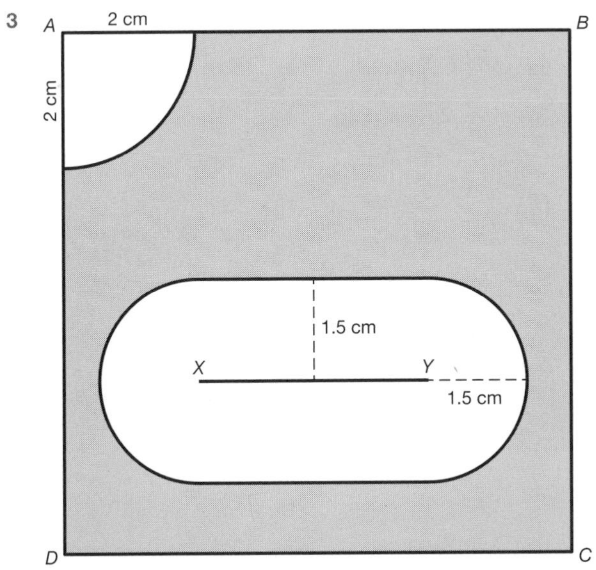

4

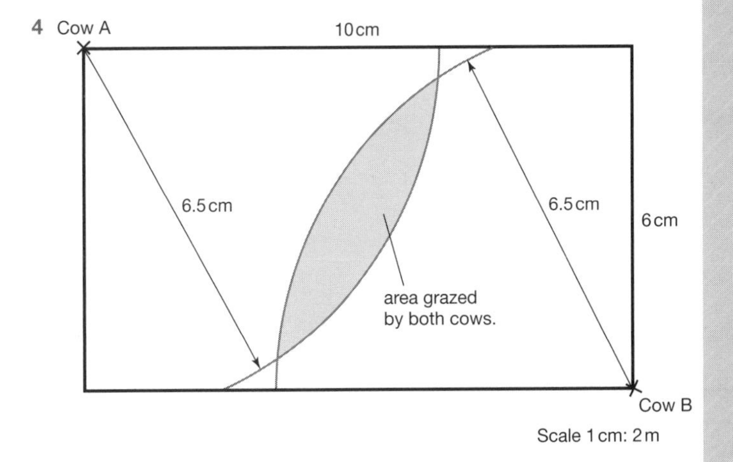

Perimeter

1 54 cm
2 11.4 cm
3 Perimeter of cushion $= \frac{1}{2} \times 2 \times \pi \times 24 + 30 + 48 + 30$
$= 183$ cm (to nearest cm)
$= 1.83$ metres. No, Greta does not have enough lace.

Area

1 a $72\,cm^2$ b $22\,cm^2$ c $82.5\,cm^2$
2 $12\,cm^2$
3 $2.63\,cm^2$

Sectors

1 a $\frac{1}{3}$ b $3\pi\,cm^2$
2 a $6.2\,cm^2$ b 10.0 cm
3 a $8.73\,cm^2$ b 3.49 cm

3D shapes

1 a

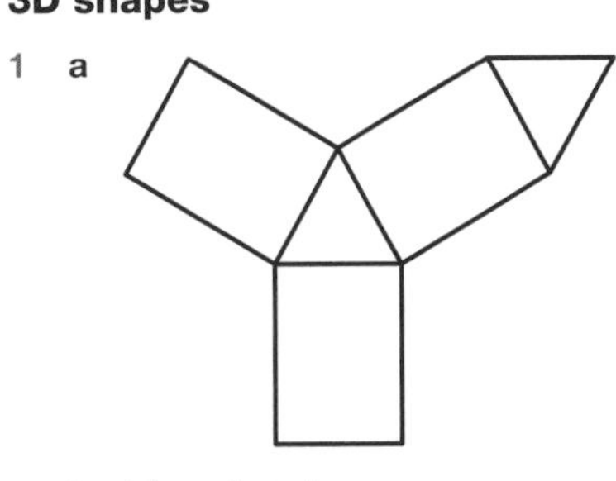

b triangular prism

c

Number of faces	Number of edges	Number of vertices
5	9	6

2 a

b

c

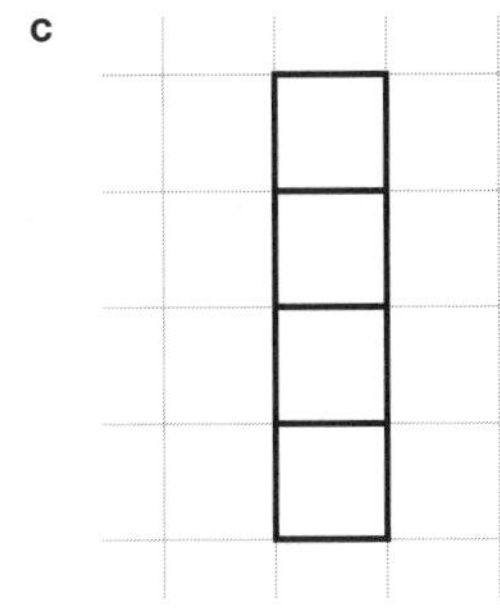

3 **a**

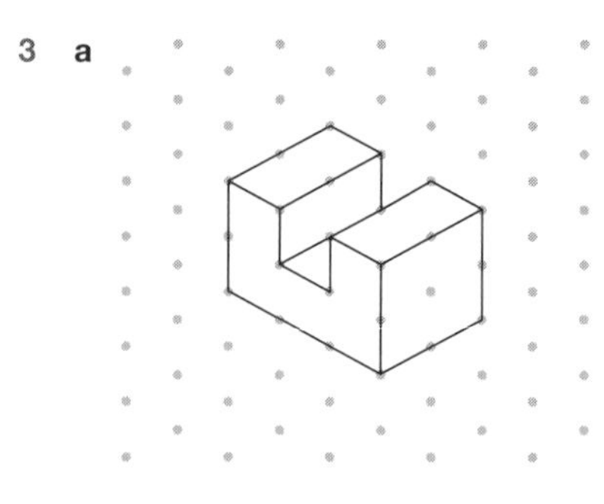

b 10

Volume

1 $20\,cm^3$

2 **a** $942\,cm^3$ **b** $770\,cm^3$

3 $1072.33\,m^3$

4 Volume of tank $= 40 \times 40 \times 60 = 96\,000\,cm^3$

Volume of water in tank, 80% full $= 0.8 \times 96\,000 = 76\,800\,cm^3$

Height of water in pond (1st fill) $= 76\,800 \div (80 \times 60) = 16\,cm$

Height of water in pond (2nd fill) $= 16 \times 2 = 32\,cm$

Height of water in pond (3rd fill) $= 16 \times 3 = 48\,cm$

Three tanks of water are needed to fill the pond.

Alternative method: divide volume of pond by volume of water in tank.

$\frac{80 \times 60 \times 48}{76\,800} = 3$

Surface area

1 **a** 6 **b** $136\,cm^2$

2 $96\,cm^2$

3 **a** $2463.01\,cm^2$ **b** $301.59\,cm^2$

4 $386.42\,cm^2$

Using Pythagoras' theorem

1 $x = 5\,cm$; $y = 9\,cm$

2 $17.9\,m^2$

3 $2\sqrt{5}$

4 diagonal of doorway$^2 = 70^2 + 190^2 =$

$\sqrt{41\,000} = 202.48\,cm = 2.0248\,m = 2.02\,m$ (2 d.p.)

Yes, the artwork will fit through the diagonal of the doorway.

Trigonometry

1 31.6° 2 22.9 cm 3 4.53 m

Exact trigonometric values

1 **a** 1 **b** 45°

2 2 cm

3 10 cm

4 45° and 45°

Transformations

1

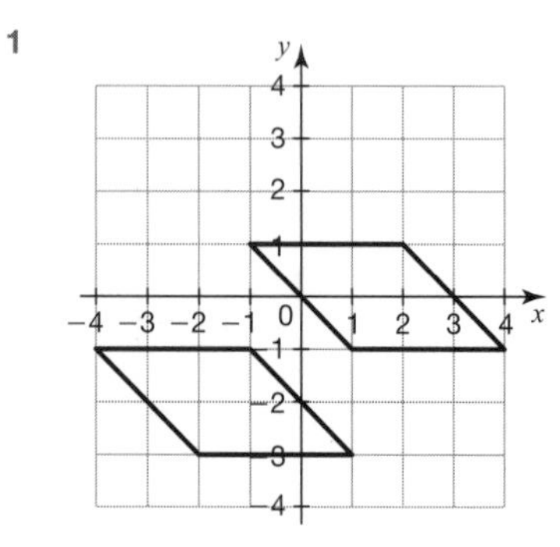

2

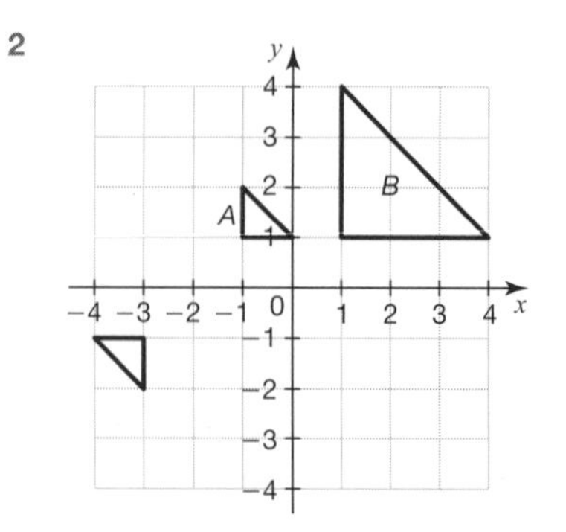

3 Rotation 90° clockwise about (1, −1); or rotation 270° anticlockwise about (1, −1).

4 **a** and **b**

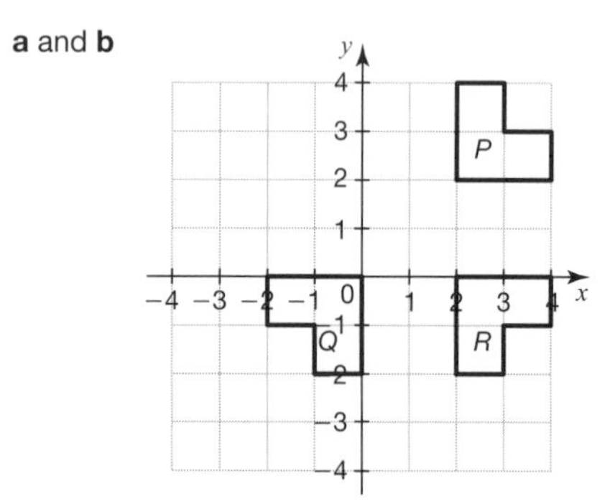

c Reflection in $y = 1$

Similar shapes

1 **a** *YZW* **b** 3 **c** 12 cm

2 **a** 37.5° **b** 4 cm **c** isosceles **d** 2.5 cm

3 **a** $\frac{2}{3}$ **b** 3

4 **a** $\frac{1}{2}$ **b** 5.6 cm **c** 46°

Vectors

1 $\mathbf{a} = \begin{pmatrix} 3 \\ 2 \end{pmatrix}$ $\mathbf{b} = \begin{pmatrix} -3 \\ 3 \end{pmatrix}$ $\mathbf{c} = \begin{pmatrix} 2 \\ -4 \end{pmatrix}$ $\mathbf{d} = \begin{pmatrix} -4 \\ -2 \end{pmatrix}$

2 **a**

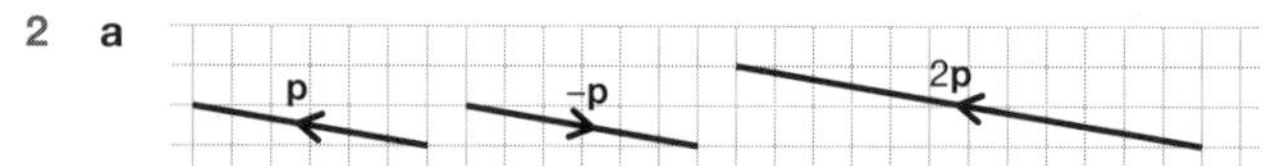

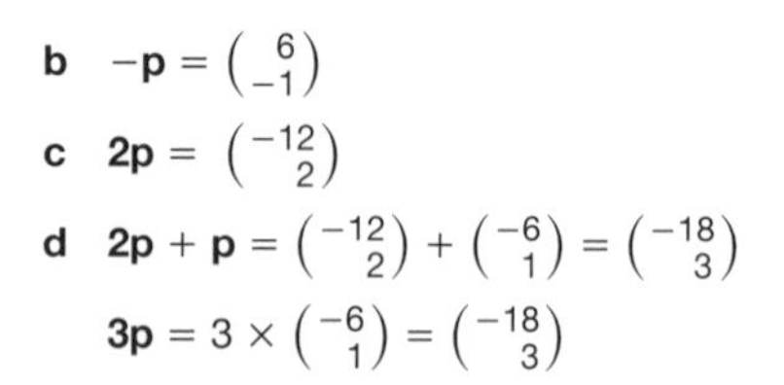

b $-\mathbf{p} = \begin{pmatrix} 6 \\ -1 \end{pmatrix}$

c $2\mathbf{p} = \begin{pmatrix} -12 \\ 2 \end{pmatrix}$

d $2\mathbf{p} + \mathbf{p} = \begin{pmatrix} -12 \\ 2 \end{pmatrix} + \begin{pmatrix} -6 \\ 1 \end{pmatrix} = \begin{pmatrix} -18 \\ 3 \end{pmatrix}$

$3\mathbf{p} = 3 \times \begin{pmatrix} -6 \\ 1 \end{pmatrix} = \begin{pmatrix} -18 \\ 3 \end{pmatrix}$

Therefore, $2\mathbf{p} + \mathbf{p} = 3\mathbf{p}$

3 **a** $\begin{pmatrix} -2 \\ 5 \end{pmatrix}$

b $\begin{pmatrix} 8 \\ 3 \end{pmatrix}$

c

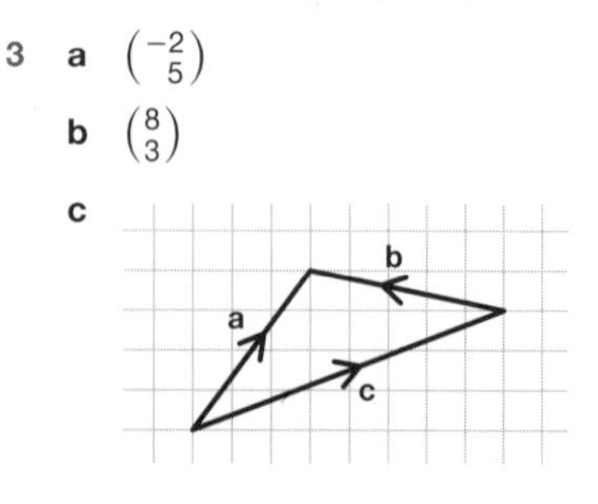

Probability

Basic probability

1 $\frac{1}{10}$ or 0.1 or 10%
2 0.4
3 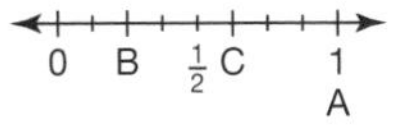

4 $2p - 0.1 + 2p + 0.1 + p = 1$
$5p = 1$
$p = 0.2$

Outcome	Red	Blue	Green
Probability	$2p - 0.1$ $= 2 \times 0.2 - 0.1$ $= 0.3$	$2p + 0.1$ $= 2 \times 0.2 + 0.1$ $= 0.5$	$p = 0.2$

Blue is most likely.

Two-way tables and sample space diagrams

1

	Single	Double	King	Totals
Oak	2	16	12	30
Pine	23	14	17	54
Walnut	1	12	3	16
Totals	26	42	32	100

2 a

		Spinner			
		1	2	3	4
Coin	Heads	1, H	2, H	3, H	4, H
	Tails	1, T	2, T	3, T	4, T

b $\frac{1}{8}$
c $\frac{3}{8}$

3 a

	Study sciences	Do not study sciences	Totals
Boys	24	21	45
Girls	35	40	75
Totals	59	61	120

b $\frac{21}{120}$
c $\frac{7}{15}$

Sets and Venn diagrams

1 a ξ = {21, 22, 23, 24, 25, 26, 27, 28, 29}
b A = {21, 24, 27}
c B = {24, 28}
d A ∪ B ={21, 24, 27, 28}
e A ∩ B ={24}

2 a 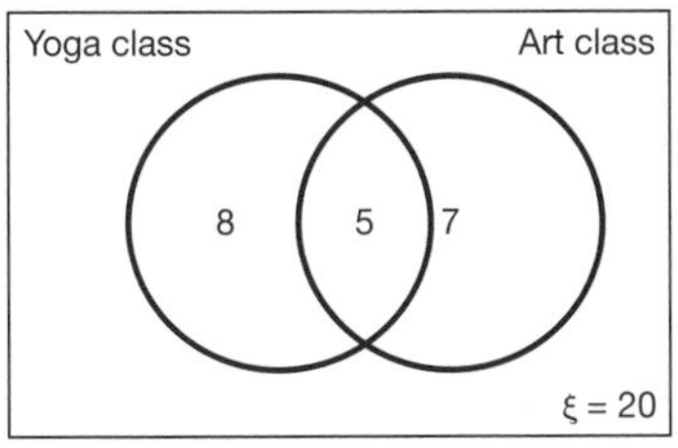

b $\frac{7}{20}$
c $\frac{3}{4}$

3 a $\frac{5}{12}$ b $\frac{2}{5}$ c $\frac{3}{20}$ d $\frac{1}{6}$

Frequency trees and tree diagrams

1 a 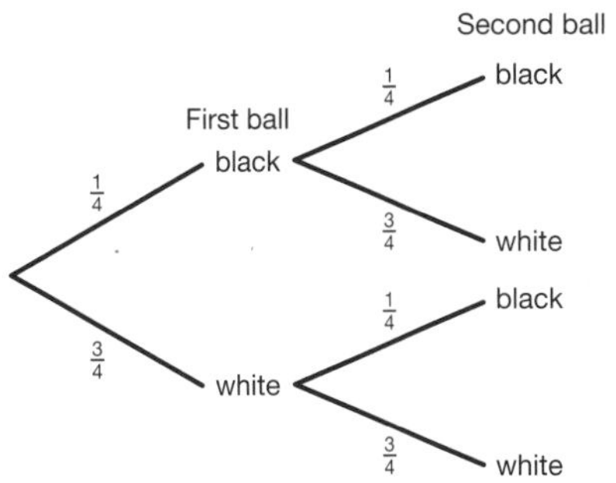

b $P(W, B) = \frac{3}{16}$
c $P(B, B) = \frac{1}{16}$

2 a 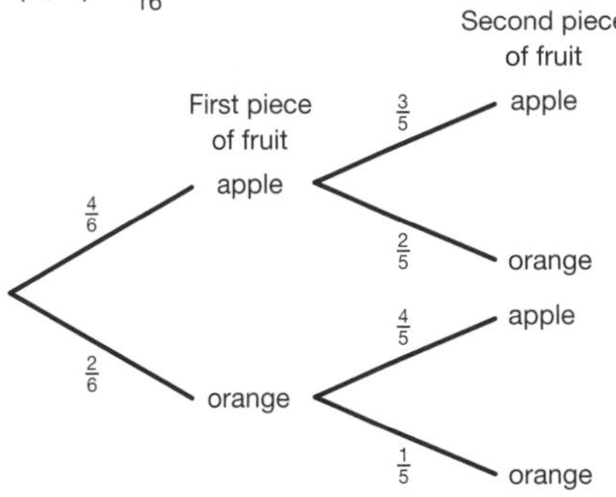

b $P(A, O) = \frac{4}{15}$
c $P(O, O) = \frac{1}{15}$
d $P(O, O)$ or $P(A, A) = \frac{7}{15}$

3 a 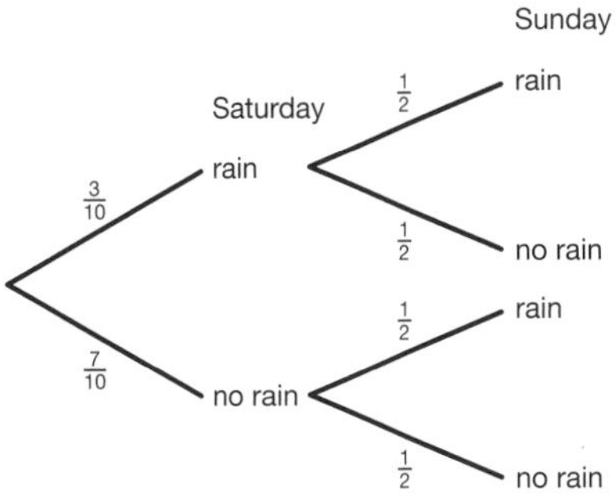

b $\frac{3}{20}$
c 50%

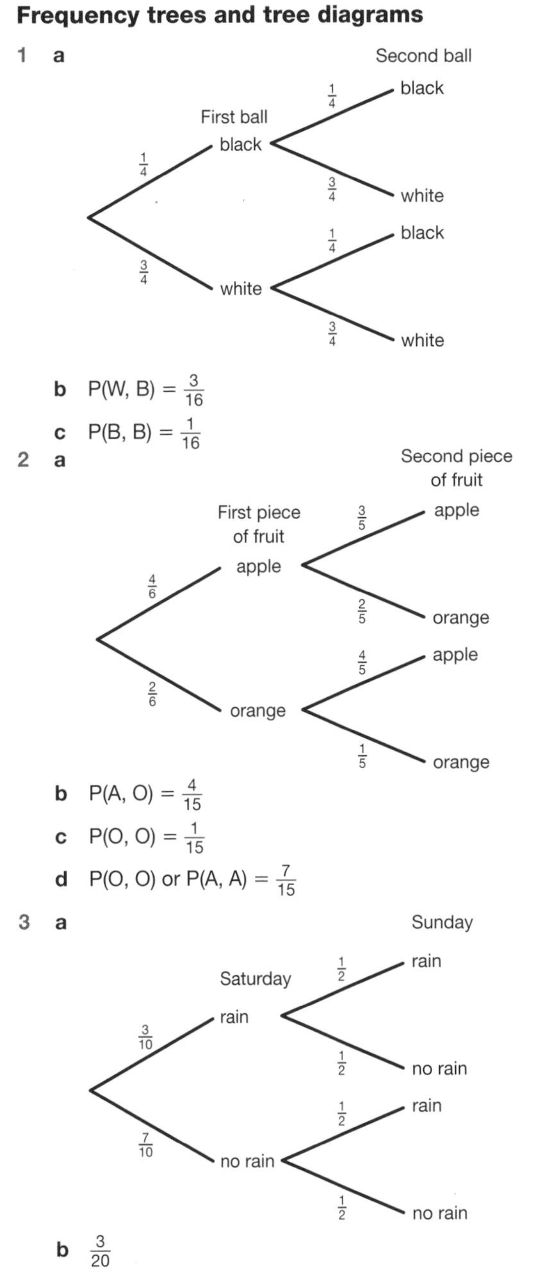

Expected outcomes and experimental probability

1 a 50 b $\frac{13}{50}$ c $\frac{3}{10}$ d 20
2 15
3 a $\frac{1}{3}$ b 150 c 45 d 225

Statistics

Data and sampling

1 65, because that is 10% of 650 (the entire population).
2 $\frac{50}{25\,000} \times 100 = 0.2\%$. The sample is not big enough.
People in the town centre may not be the only ones using buses. For example, some people may take buses to the local train station, school or hospital.
3 a 80 whole cakes
b Assumptions
If answered 80, then assuming each person will eat a whole cake; it could affect the answer if instead assume each cake is divided into 8 slices (e.g. then only 10 carrot cakes would be needed).
Assumed the sample is representative of the population; this could affect the answer because not all the 400 people who have accepted the invitation may turn up.

Frequency tables

1 a 30 b 91

2

Number of electronic devices	Tally	Frequency
0–1	|||	3
2–3	~~||||~~ ~~||||~~	10
4–5	~~||||~~	5
6–7	~~||||~~	5
8–9	|	1

3 a Continuous

b

Mass, m (kg)	Tally	Frequency
$50 \leq m < 60$	|||	3
$60 \leq m < 70$	~~||||~~	5
$70 \leq m < 80$	||||	4
$80 \leq m < 90$	~~||||~~	5
$90 \leq m < 100$	|||	3

Bar charts and pictograms

1 a 5 b 41

2

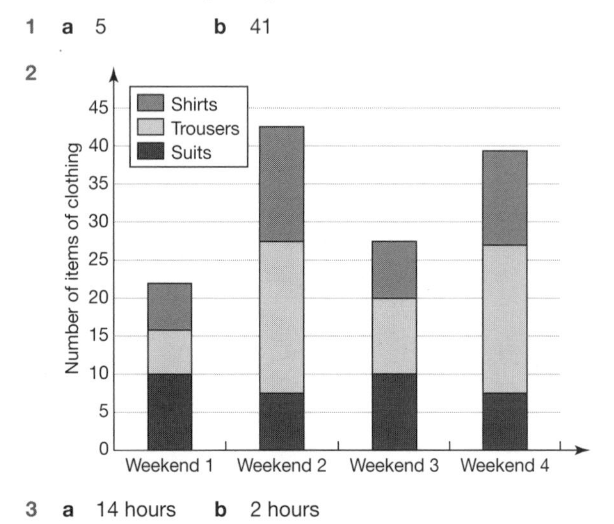

3 a 14 hours b 2 hours

Pie charts

1 a 125 b 225

2

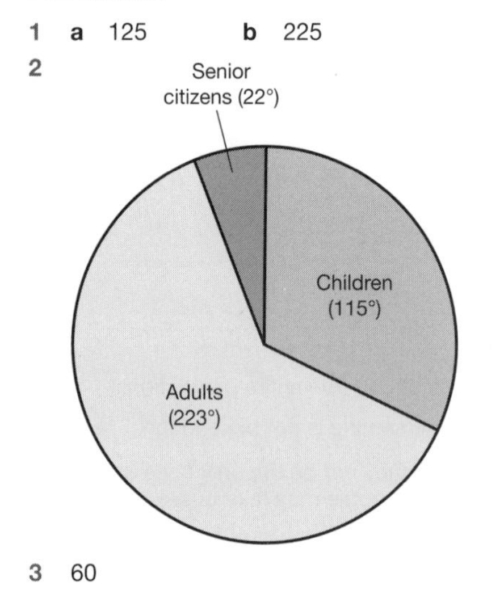

3 60

Stem and leaf diagrams

1 a 71 cm b 20

2 a 2.23 m b 3

3

Year 9s		Year 10s
95 80 50	0	65 75
75 30	1	50 85
65 00	2	70
05	3	10

Key
Year 9s 50|0 means £0.50
Year 10s 0|65 means £0.65

Measures of central tendency: mode

1 3 minutes and 4 minutes
2 $12 < a \leq 13$
3 £10–£20
4 25 kg

Measures of central tendency: median

1 $14\frac{1}{2}$
2 $12 < a \leq 13$
3 $28\frac{1}{2}$

Measures of central tendency: mean

1 11
2 3 bedrooms (rounded to the nearest whole number)
3 1 holiday (rounded to the nearest whole number)
4 46 years old (rounded to the nearest year)

Range

1 a 2 years b 3 years
2 a 9°C b 15°C
3 a Business A: range = £22 255; mean = £34 768
b Business B: range = £45 354; mean = £38 572.50
c Either: Business A because its range in profit is lower and the profit is increasing each year, and so it shows a more consistent performance. OR: Business B because its mean profit is higher, and its most recent profit (in Year 4) is £17 432 more than Business A.

Comparing data using measures of central tendency and range

1 a 34.5 minutes
b 13 minutes
c 28.5 minutes
d 54 minutes
e Either: The bus is better because although it takes longer (on average), the range is lower, and so you can predict the time it takes for the journey.
OR: The train is better because it is quicker than the bus (on average), although the range suggests it may be less reliable.

2 a Mean = 13. This does not represent the age of the people using the playground. In fact, the ages of those using the playground are small children (under 10) and their parents (over 25).
b There are five modes (3, 4, 5, 7, 8), and so the mode does not represent the age of the people using the playground.
c Median = 7

3 Mode = 0; Median = 0; Mean = 2 days. Mode or median are the best averages to use, because the mean is skewed by the student who is absent due to sickness for 24 days.

Time series graphs

1 a 190 g b 5 days c 20 g

2 a

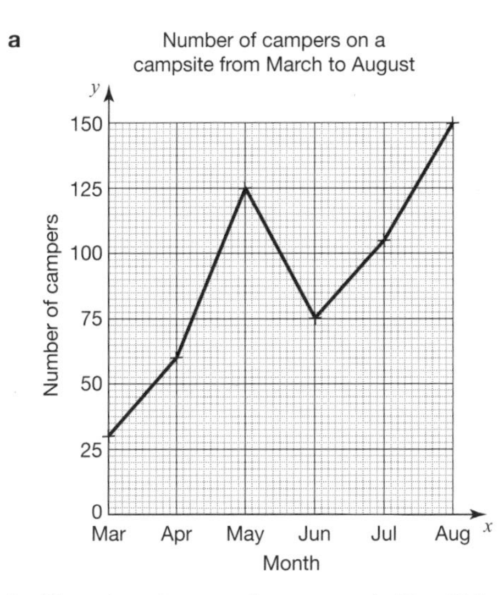

b There is an increase in campers in May. This may be due to May bank holidays, or May half-term, or perhaps there was some very sunny weather.

3 a

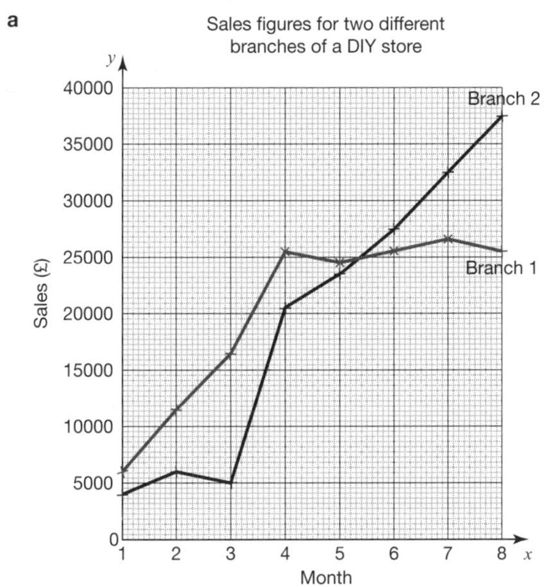

b Branch 1 had a steady increase in sales for the first four months. Then sales levelled off to stay at around £25 000. Branch 2 had a slow start to its sales in the first three months. Then perhaps it had a promotion, because sales increased a lot in month 4. Sales have been increasing ever since.

Scatter graphs

1 a Positive correlation. This means as the temperature rises, more pairs of flip flops are sold.

b Negative correlation. This means as the temperature rises, fewer wellington boots are sold.

2 a

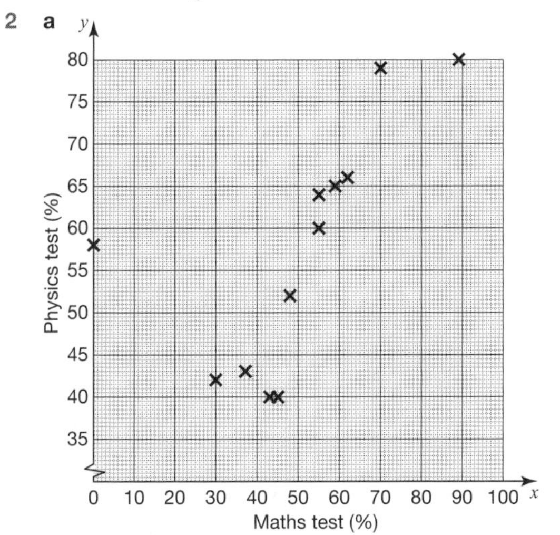

b The scatter diagram shows a positive correlation between students' maths and physics test percentages. Therefore, the students who got a low percentage in the maths test got the lower percentages in the physics test; the students who got a high percentage in the maths test got the higher percentages in the physics test.

c The outlier is the point marked at (0, 58)

d The student was absent for the maths test.

3 a

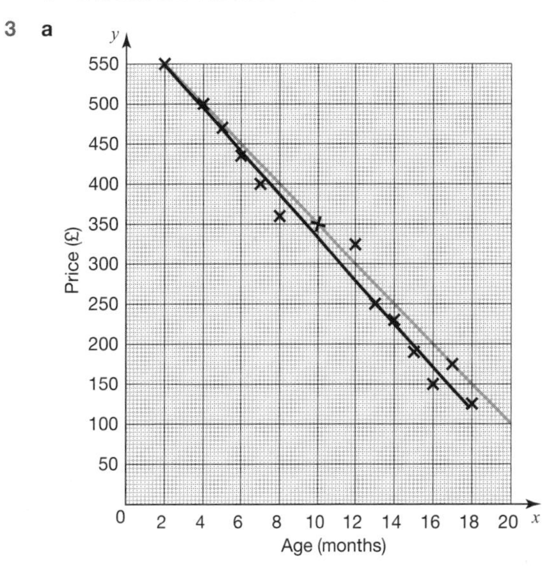

The black line shows the line of best fit.

The grey line shows the line where a laptop loses £150 every 6 months.

The shop owner is not correct. The line of best fit shows on average a laptop loses approximately £159/£160 every 6 months.

b The line of best fit cannot make a prediction outside the available data. The data only goes as far as 18 months.

Practice paper

Non-calculator

1 7000

2 20

3 No, Sandeep is not correct:

$\frac{2}{5} = \frac{4}{10} = 0.4$

But 4% = 0.04

4 3 and 5

5 120 m

6 $\frac{2}{5}$

7 Yes. Fun run + music festival = £9689 + £9689 + £6370 = £25 748

8 $\frac{7}{10}$

9 Prize A = 8 × 4 = 32 tickets

Prize B = 2 + 4 + 8 + 16 = 30 tickets

Prize A gives more tickets

10 a 30

b No, Harry is incorrect. The number of triangles is not the pattern number multiplied by 4. Rather, it is add 4 triangles each time.

11 a £3.50 b £5.00

12 £180

13 11 am

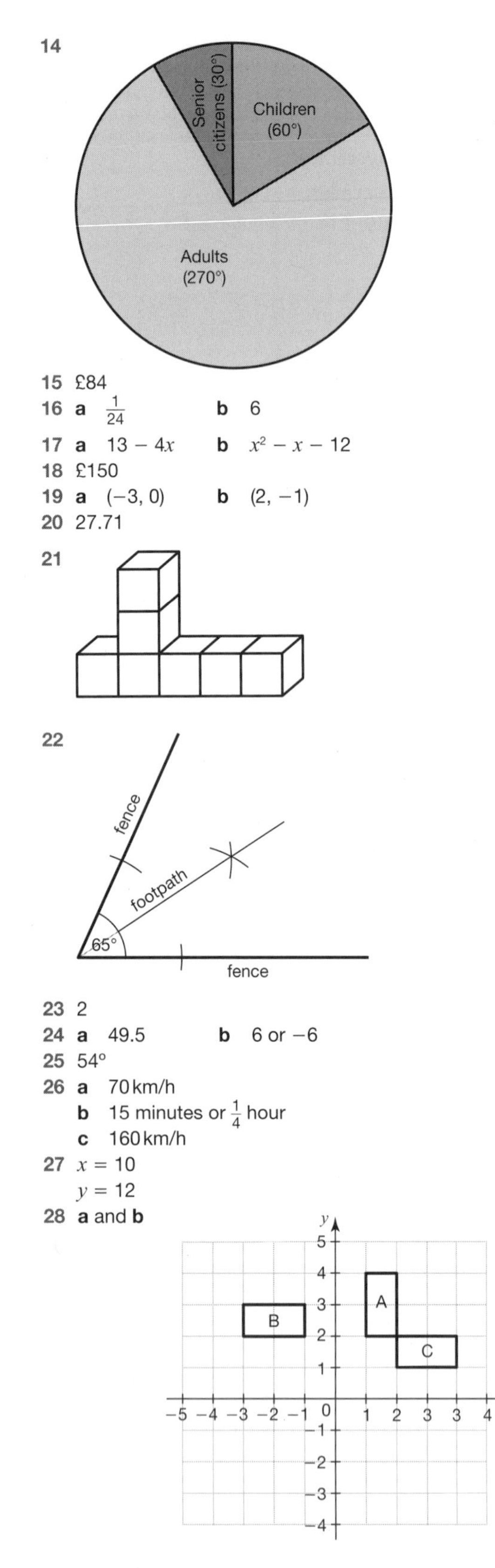

14

15 £84

16 **a** $\frac{1}{24}$ **b** 6

17 **a** $13 - 4x$ **b** $x^2 - x - 12$

18 £150

19 **a** (−3, 0) **b** (2, −1)

20 27.71

21

22

23 2

24 **a** 49.5 **b** 6 or −6

25 54°

26 **a** 70 km/h
b 15 minutes or $\frac{1}{4}$ hour
c 160 km/h

27 $x = 10$
$y = 12$

28 **a** and **b**

c Reflection in $x = y$, or rotation of 90° anticlockwise about the point (1.5, 1.5).

29 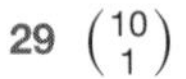 $\begin{pmatrix} 10 \\ 1 \end{pmatrix}$

Practice paper

Calculator

1 0.875
2 £490
3 £619.75
4 £259 = €295.80. The games console is cheaper in the UK.
5 19.4
6 2.4 cm
7 **a** Friday **b** 10% **c** $\frac{5}{7}$
8 1 and 81; 3 and 27
9 −1, 0, 1, 2, 3, 4
10 £67.50
11 546 cm³
12 23.3 m/s
13 £60
14 $2\sqrt{5}$
15 126
16 5.8×10^7 km
17 $3x + 8$
18 **a**

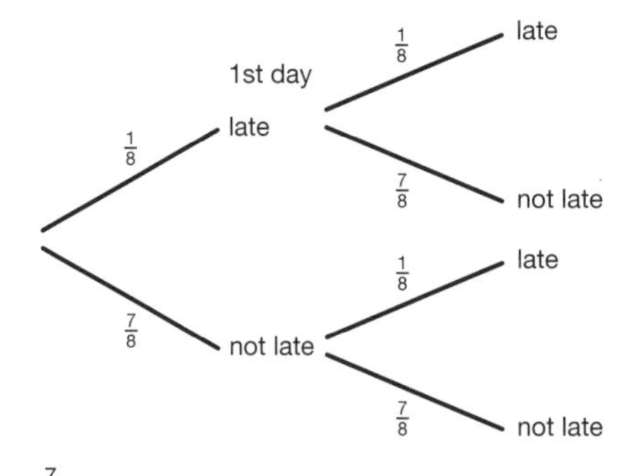

b 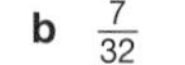 $\frac{7}{32}$

19 $x = 58°$; $y = 5.67$ cm
20 Team A average = 2; Team B average = 1.8; Team A scored more goals, on average.
21 **a** Option A is best value if you plan to visit the gym once per week, as in one month 4 visits cost £20 (or 5 visits cost £25). This is less than the monthly fee of £30.
b 6
22 58.9 m²
23 **a**

x	−2	−1	0	1	2
y	5	−1	−3	−1	5

b

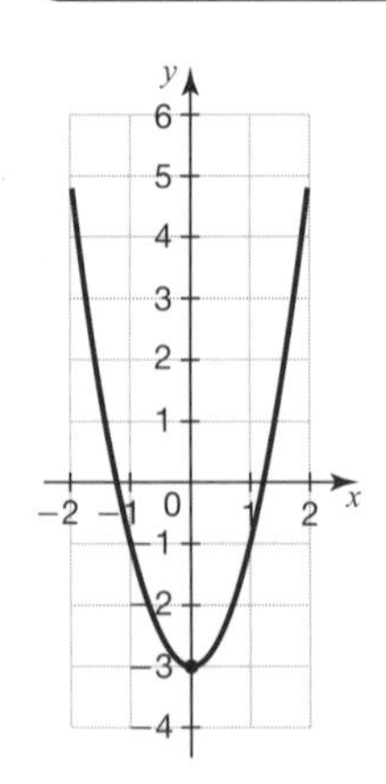

24 **a** You should draw a line of best fit to show, yes, the car dealer is correct.
b Any answer from £3500 to £4000 is acceptable.
25 $n^2 + 2n - 15$